Alfred Lambremont Webre's *The Dimensional Ecology of the Omniverse* is of extreme importance because multidimensional intelligence opened to all humans when the Mayan Calendar ended in 2011/2012. We are receiving contact from the fully inhabited omniverse that connects with us by telepathy and teleportation in the quantum dimension of our world. Webre's hypothesis— *dimensionality is the key design criteria* of the Omniverse as discrete bands of conscious energy by which intelligence organizes itself—is synchronous with my nine-dimensional model that describes how other dimensions contact humans.

Webre's compact and easy-to-read book is a brilliant compilation of solid scientific evidence for the existence of human souls that reincarnate and exist in parallel lives in many dimensions, frequently as extraterrestrials! This is a must-read for anyone who wants to fully engage with many dimensions to end absolutely any fear of abandonment, death, or the future.

— BARBARA HAND CLOW

CO-AUTHOR WITH GERRY CLOW OF *ALCHEMY OF NINE DIMENSIONS: THE 2011/2012 PROPHECIES AND NINE DIMENSIONS OF CONSCIOUSNESS.*

THE DIMENSIONAL ECOLOGY OF THE
OMNIVERSE

The Dimensional Ecology of the Omniverse

*How God and Souls in the Afterlife Create and
Inhabit Intelligent Civilizations in Our Multiverse*

Alfred Lambremont Webre, J.D., M.Ed.
Edited by Aidan Kelly, PhD

Copyright © 2014 by Geri DeStefano-Webre, Ph.D., and Alfred Lambremont Webre, J.D., M.Ed.

All rights reserved. No part of this publication may be reproduced, stored in a retrieval system or transmitted, in any form or by any means, electronic, mechanical, photocopying, recording, or otherwise, without the prior written permission of the author except for the use of brief quotations in which the source of the quotation is referenced.

For copyright licenses, please contact: *editor@universebooks.com*
Universe Books *www.universebooks.com*

Typesetting and Cover Design by *www.wordzworth.com*

Library and Archives Canada Cataloguing in Publication
Webre, Alfred Lambremont, 1942-

Library of Congress Cataloguing

Life After Death, God, Reincarnation
Intelligent Civilizations in the Omniverse
Alfred Lambremont Webre

ISBN: 978-0-9737663-1-8

1. God 2. Life After Death 3. Reincarnation 4. Soul 5. Omniverse 6. Outer space 5. Intelligent Life 6. Science

Other Books by Alfred Lambremont Webre

Exopolitics: Politics, Government and Law in the Universe
The Levesque Cases
The Age of Cataclysm

Contents

Acknowledgments — iii
Introduction: A Map of the Omniverse — v

PART I   THE OMNIVERSE AND ITS INHABITANTS — 1

Chapter 1   The Omniverse — 3
Chapter 2   Typology of Intelligent Civilizations — 33

PART II   EXOPOLITICAL DIMENSIONS OF THE OMNIVERSE — 49

Chapter 3   Time-Space Solar-System Civilizations — 51
Chapter 4   Hyperdimensional Civilizations — 81
Chapter 5   Intelligent Civilizations' Governance Authorities — 101

PART III   SPIRITUAL DIMENSIONS OF THE OMNIVERSE — 121

Chapter 6   The Interlife Dimensions — 123
Chapter 7   The Intelligent Civilizations of Souls — 143
Chapter 8   Hyperdimensional Civilizations' Role in Dimensional Ecology — 163
Chapter 9   The Science of the Omniverse — 175

Sources and Resources — 185
Index — 207
About author Alfred Lambremont Webre — 215

Acknowledgments

Given that this is a book on the dimensional ecology of the Omniverse, I would like to acknowledge the invaluable support and insights I have received in this lifetime from those dimensional persons and forces in Exopolitical and Spiritual dimensions that have led to the writing of this book. I would like to thank my guides, assisting hyperdimensional civilizations, assisting spiritual entities, and assisting souls, as well as the Source. Thanks also to those that have attempted to prevent the publication of this book, for you have only made it more refined.

I thank and acknowledge the incalculable help I have received over the years in the Earth dimensions from many colleagues, researchers, and friends too numerous to mention here.

I thank my editor, Aidan Kelly, who helped shape this book into a vessel for you, the reader, as he has done with so many books before as an editor with Scientific American and Jeremy Tarcher (Penguin).

I thank my families, who bore patiently my years of research into the dimensional ecology, including my parents, my siblings, my son, and my step-daughters, and my soul spouse, Geri DeStefano-Webre, Ph.D., without whose presence this book would not have been possible.

And I thank you, the reader, who now hold yet another map into the dimensional ecology of the Omniverse in your hands. May it serve you and yours well.

Introduction:
A Map of the Omniverse

Welcome to a map of the dimensional ecology of the Omniverse. Our map of the Omniverse is based on replicable, empirical *prima facie* evidence that is evaluated according to the law of evidence. The map establishes the existence of a dimensional ecology of intelligent civilizations in the physical universes of time, space, energy, and matter in our multiverse. The dimensional ecology includes the intelligent civilizations of souls, spiritual beings, and Source (God) in the spiritual dimensions. The Omniverse, this *prima facie* evidence establishes, is the sum of the multiverse plus the spiritual dimensions—in other words, all that is.

As I was mapping the Omniverse for this book, a reader emailed me a question that caused me to look more deeply into the reasons why I was writing a book that applied the law of evidence to demonstrate the dimensional ecology of the Omniverse. He wrote:

> I do have a somewhat personal question for you. I am a spiritual person and not religious at all, although I am personally a Christian by faith. I have been listening to you for some time. I first found you back when dial-up connections were the only way to access the internet, but in all that time I do not remember if I ever heard you talk about your personal faith or not. So my curiosity is, are you a spiritual person? Do you believe in any particular faith?
>
> Thank you for your years of work in public service.

After some thought, I wrote back,

I am very happy you have asked that question. A number of people have asked me the same question recently, so it feels as though I should address this issue publicly.

1. Spiritual history. I am a very spiritual person and was raised in religious boarding schools from a young age where prayer and meditation were encouraged. As an adult, prayer and meditation have formed a large part of everyday life. I do not ascribe to a formal "religion" as such. I left being a Roman Catholic in my early 20s while at University. I feel a direct connection to Source and to the process of Creation and spiritual reality in all there is. I have had direct experiences with the Divine that are highly personal that I will be writing about in a future book (not now). This is my private personal life.

2. Public expression, my public expression and *persona*. Because of my training in law, journalism, and Exopolitics, I do not write or speak about 'God' and spirituality directly except insofar as I can provide empirical evidence of entities and realities such as Source, souls, spiritual beings, and spiritual dimensions.

The closest that I have gotten to speaking about God and spirituality is in public lectures at conferences that I deliver on "The Dimensional Ecology of the Omniverse." These lectures are based on data whose validity I can prove according the law of evidence. My lectures advance concepts based on evidence, not on personal belief.

My lectures explore the empirical evidence for the existence of an original Source (God) and of souls and of an afterlife. The lectures explore an ecology of interlocking dimensions of existence among the universes of the multiverse, the afterlife, and the spiritual dimensions or what we term the "Omniverse."

I do not publish my personal beliefs, only concepts I can prove with replicable empirical evidence or at least *prima facie* evidence. Hence to date I have not published using the terms God, souls, spirit, afterlife, etc. An early summary of my research appeared in "Human souls common to extraterrestrials and earthlings, researcher says."[1]

---

[1] Webre, "Human souls common," http://bit.ly/18mlpLv.

INTRODUCTION: A MAP OF THE OMNIVERSE

My next book, *The Dimensional Ecology of the Omniverse*, explores the empirical evidence for a dimensional ecology that includes not only Earthling humans, extraterrestrials, and hyperdimensionals in our universe and in the multiverse, but also the empirical evidence that this dimensional ecology extends to souls, the afterlife, spiritual beings, and Source (God) in the spiritual dimensions. Together, the physical universes of the multiverse plus the unseen dimensions of the spiritual dimensions form the Omniverse, or all that is.

You will find that with this new platform of empirical evidence, I will be speaking and writing publicly on the nature of the dimensional ecology of souls, spirit, and God (Source), starting with the publication of my book on the Omniverse.

SPAIN AND THE DIMENSIONAL ECOLOGY

My journey in mapping the dimensional ecology of the Omniverse experienced a significant threshold in Barcelona, Spain. On a bright day in July 2009 in Barcelona, Ima Sanchis, a veteran journalist with *La Vanguardia*, the leading Catalan newspaper and top newspaper in Spain, interviewed me for her full-page column *La Contra*. I was on a book tour of Spain as part of the launch of the Spanish edition of my book *Exopolitics: Politics, Government, and Law in the Universe.*"[2]

Toward the end of the interview, in a small conference room at the La *Vanguardia*'s Barcelona headquarters, Ima put down her pen, turned off her tape recorder, turned to me, and pointedly asked,

"¿Alfred, Como funciona el universo? ¿Cuál es la relación entre los extraterrestres y la reencarnación del alma humano?"

Translated into English, Ima Sanchis' questions mean,

"Alfred, how does the universe function? What is the relationship between extraterrestrials and the reincarnation of the human soul?"

---

[2] The Spanish edition is Exopolitica: La Politica, El Gobierno y La Ley en el Universo (Vesica Piscis, 2009), http://amzn.to/185GYdq.

As Ima asked her questions, I recognized that she was asking me to address scientifically and on the record those key issues that most humans, philosophers, scientists, priests, and earthly religions have struggled to answer for millennia–without much objective scientific success. It dawned on me in that moment that Ima was pointing to a next pathway of empirical research into how intelligent civilizations in the physical multiverse relate to each other and to the larger community of intelligent civilizations of souls, spiritual beings, and Source. Now was the time to integrate the work in parapsychology and extraterrestrial studies I began in 1973.

I returned to Madrid from speaking in Barcelona. At my hotel in Madrid, on my way back home to Vancouver, B.C., I pondered the relationships between extraterrestrials and the soul, and scribbled notes to myself on hotel notepaper. My mind explored the basic similarities in communications and transportation between humans and extraterrestrials in this universe and between humans and souls in the "Interlife." There were structural parallels between those two fields of inquiry in how information about them is organized—yet no one, as far as I knew, had ever looked explicitly at those patterns.

Our research has found replicable *prima facie* evidence for the existence of reincarnation as a regular feature of intelligent life in the universes of the multiverse. Hence we use the term "Interlife" interchangeably with "afterlife" to designate the period during which a soul locates fully in the spiritual dimensions between incarnations in a universe in the Exopolitical dimensions of the Omniverse.

TELEPATHY AND TELEPORTATION

Telepathy and teleportation are two common modalities by which intelligent civilizations navigate the dimensional ecology of the Omniverse. Communications among human contactees and representatives of hyperdimensional intelligent civilizations can be telepathic, as exemplified by the reported extraterrestrial contactee cases. A common mode of transportation among human contactees and representatives of intelligent civilizations can be teleportation, as when human abductees are teleported to a waiting spacecraft by a hyper-dimensional civilization.

Communications between mediums and representatives of "soul civilizations" in the afterlife (or the "'Interlife"), as exemplified by the many apparently genuine reported cases of mediumistic communications with persons in the afterlife, are also telepathic. Transportation between the time-space dimension in our physical universe and the afterlife also appears to be teleportation. For example, parapsychological research reports that when an individual soul departs from a physical body that has died, the soul teleports to an interdimensional portal between our time-space dimension and the afterlife dimension.

These common factors—telepathy (nonlocal mind-to-mind communication) and teleportation (displacement between points in a single dimension or between points among several dimensions)—seem to reveal fundamental similarities between communication and transportation among humans, intelligent civilizations of extraterrestrials, and intelligent civilizations of souls.

Dimension-Based Typology of Intelligent Civilizations

A second important threshold along the mapping of the dimensional ecology of the Omniverse occurred shortly afterward, in August 2009, when an editor for Oxford University Press asked me to write a book applying "the principles of international law to extraterrestrial contact."

In order to write a book on that subject, I realized I would have to first define what the term "intelligent extraterrestrial civilizations" means—and at this point I discovered that no adequate typology of such civilizations had yet been developed. The existing typologies did not accurately reflect the way intelligent civilizations self-report themselves, as being based in a specific dimension or energy frequency. The specific dimension (or "density") in the dimensional ecology of consciousness in which an intelligent civilization is based is that civilization's most fundamental typological, locational, and developmental criterion.

That line of thought led to development of a dimension-based typology of intelligent civilizations, based on recent whistleblower, direct witness, and documentary evidence. This new typology establishes dimension-based criteria for extraterrestrial and interdimensional civilizations and extraterrestrial governance bodies. As we shall see, the dimension-based typology of intelligent extraterrestrial civilizations in turn provides an evidentiary and conceptual foundation for mapping the dimensional ecology of the Omniverse.

THE OMNIVERSE

Our mapping of the Omniverse starts in earnest in **Chapter 1** where, in the words of my editor (who also edited books by the distinguished late astronomer Sir Fred Hoyle of Cambridge University), we are thrown "right into the deep end of the pool."

Exopolitical dimensions

Our map provides an overview of the dimensional ecology of the Omniverse through which the intelligent civilizations of souls, spiritual beings, and Source (God) in the spiritual dimensions co-create and inhabit the intelligent civilizations in the Exopolitical dimensions of the universes of time, space, matter, and energy in the multiverse.

INTRODUCTION: A MAP OF THE OMNIVERSE

The universes of the multiverse are also collectively termed the "Exopolitical dimensions" to highlight the function of the physical multiverse as a base for intelligent extraterrestrial and hyperdimensional civilizations. The science of Exopolitics, which studies relations between intelligent civilizations in the multiverse, is the science of the dimensional ecology in the universes of the multiverse.

Spiritual dimensions

Our map reviews concepts of our universe and the multiverse, and explores current scientific estimates of how many universes exist in our multiverse. Our map reviews the best empirically based estimates of the number of intelligent civilizations in our galaxy, universe, and multiverse. It also introduces empirical concepts of the intelligent civilizations of souls, of spiritual beings, and of Source (God) in the spiritual dimensions.

Why is this important? Because this data provides a graphic picture of how "humongously" [in the words of one scientist] large the multiverse and Omniverse are, in order to provide a sense of perspective on ourselves as an intelligent planetary civilization in our universe.

The Omniverse provides what string theory does not

The dimensional ecology of the Omniverse provides what "string theory" has not been able to, and expands on the conventional scientific definition of the multiverse as consisting solely of parallel physical universes of time, space, energy, and matter, our physical universe being one of them. The spiritual dimension of the Omniverse provides the energy that human scientists cannot now account for in the creation and maintenance of each physical universe in the multiverse.

Dimensional Ecology: Equations

Our mapping reviews the basic equations for the dimensional ecology of the Omniverse:

Multiverse (Exopolitical dimensions)
= Sum of all physical universes

Omniverse = multiverse (Exopolitical dimensions)
+ spiritual dimensions (intelligent civilizations of souls, spiritual beings, and Source / God)

Dimensional Ecology and Religious Belief

The dimensional ecology of the Omniverse is both a scientific hypothesis and a truth that many Earthling humans are familiar with intuitively or have learned during their lives in their families, schools, religions, reading, or philosophizing. Humans are familiar with general aspects of the dimensional ecology, such as "afterlife," "souls," and "Source" (God), although they may have widely differing opinions depending on race, nationality, gender, and education as to whether the afterlife, souls, and Source (God) actually exist and in what specific form.

Many assertions about specific aspects of the dimensional ecology, such as life after death, are set out in religious beliefs and sacred texts that organized religions have relied on over millennia to win followers. Yet important details about the nature of the afterlife as reported in human religions, for example, are sometimes only partially correct or factually incorrect when compared to replicable, scientifically derived knowledge about the nature of the afterlife.

INTRODUCTION: A MAP OF THE OMNIVERSE

THE MULTIVERSE

Our map turns its focus to the multiverse, or the Exopolitical dimensions of the Omniverse. Starting in **Chapter 2** and, continuing through **Chapter 5**, it explores the dimensional ecology of the intelligent civilizations based in the physical universes of time, space, energy, and matter in the multiverse.

Chapter 2 introduces the dimension-based typology of intelligent civilizations in the multiverse, a typology based on how hyperdimensional intelligent civilizations describe themselves. The dimension-based typology distinguishes between intelligent civilizations in the time-space dimension; hyperdimensional civilizations; and intelligent civilizations' governance authorities in the universes of the multiverse.

Exophenotypology

Concurrent with the dimension in which a specific intelligent species or civilization is based, an intelligent species or civilization can also be typed by "Exophenotypology," which is the typology or classification of extraterrestrials based on the observable characteristics of their physical appearance. Our map of the dimensional ecology explores the humanoid, Grey, Reptilian, Insectoid, and other Exophenotypologies of intelligent civilizations in the multiverse.

TIME-SPACE SOLAR SYSTEM CIVILIZATIONS

Chapter 3 explores a case study of the evidence for an intelligent civilization based in the time-space dimension on a planet in our own solar system, in this case, the existence of three humanoid Exophenotypes on Mars. This conclusion is the result of empirical *prima facie* evidence and application of the law of evidence, and strongly supports the underlying hypothesis of the dimensional ecology of the Omniverse.

Our map goes on to provide an Exophenotypology of Martian humanoids, based on the *prima facie* evidence of eyewitnesses and of documentary (photographic) evidence, including that of the NASA Mars rovers that have photographed three distinct Martian Exophenotypes on the surface of Mars.

At this historical stage in the exploration of Mars, the appropriate standard of proof in determining whether the available eyewitness and documentary evidence shows that indigenous intelligent life does exist on Mars is *prima facie* evidence, or "on the face of it" evidence. There is no scientific, legal, or ethical requirement that a decision about whether intelligent life exists on Mars must satisfy the whimsical criterion proposed by Carl Sagan that "Extraordinary claims require extraordinary evidence."[3]

HYPERDIMENSIONAL CIVILIZATIONS

Chapter 4 explores the *prima facie* empirical evidence for how hyperdimensional intelligent civilizations operate within the dimensional ecology of the Exopolitical dimensions in the multiverse. Hyperdimensional civilizations are intelligent civilizations based in dimensions in our universe parallel to our own time-space Earth dimension or in universes parallel to our universe. Hyperdimensional civilizations may use technologically advanced interdimensional transport when teleporting into our known physical universe or our Earth time-space dimension. Hyperdimensional civilizations may also use advanced consciousness technologies that permit them to teleport interdimensionally.

Hyperdimensional intelligent civilizations can be based in higher dimensions of our own solar system, of our galaxy (the Milky Way) or of other galaxies of our universe, or of universes parallel to our own universe in the multiverse.

---

[3] Satoshi Kanazawa, "Do extraordinary claims require extraordinary evidence?" http://bit.ly/1chFcqK.

INTRODUCTION: A MAP OF THE OMNIVERSE

Intelligent Civilizations' Governance Authorities

Chapter 5 explores the *prima facie* evidence for the existence of intelligent civilizations' governance authorities, including types of witness and documentary evidence that is within the protocols of evidence acceptable to the science of Exopolitics.

There are at least two sources of *prima facie* Exopolitical evidence for legally constituted intelligent civilizations' governance authorities that exist in the dimensional ecology of the universes of the multiverse and that have jurisdiction over a defined dimensional territory, such as the Milky Way galaxy in our universe. The first such source is replicable scientific remote viewing. The second is eyewitness and documentary evidence, consisting of eyewitness contactee and telepathic interaction with representatives and intermediaries of reported intelligent civilizations' governance authorities.

Spiritual Dimensions of the Omniverse

Chapters 6 through 9 explore the dimensional ecology of the spiritual dimensions of the Omniverse. *Prima facie* empirical evidence supports the hypothesis that the Omniverse consists of the totality of parallel universes in the Exopolitical dimensions (the multiverse) plus the spiritual dimensions. The parallel "visible" universes (or the multiverse) can be more aptly termed the Exopolitical dimensions, since they are where the intelligent civilizations of souls that are based in the spiritual dimensions of the Omniverse incarnate in a variety of intelligent creatures for the purpose of moral and soul development.

Evidence for the existence of the spiritual dimensions and dimensional ecology between the Exopolitical and spiritual dimensions is provided by the evidence for the existence of the continuation of consciousness in the Interlife after bodily death in the Exopolitical dimensions. The communications of intelligent civilizations of souls based in the Interlife dimensions to

Earthlings in the time-space hologram via the dimensional ecology provide the evidence for the existence of souls, the Interlife, and the dimensional ecology itself.

Chapter 6 explores the *prima facie* evidence for the dimensional ecology of the Interlife (afterlife) dimensions. Our map notes that the same parties that developed Instrumental Transcommunication (ITC) as a technology for communicating in the dimensional ecology between the Interlife in the spiritual dimensions and the universes of the multiverse also developed chronovision, a time-travel technology for exploring timelines in the multiverse.

It is highly meaningful for the dimensional ecology of the Omniverse hypothesis that ITC's Electronic Voice Phenomena (EVP) for exploring the Interlife in the spiritual dimensions and chronovision have a common origin in the work of two Italian Catholic priests, Fathers Pellegrino Ernetti and Augustino Gemelli.

Chapter 7 explores the *prima facie* evidence for the dimensional ecology of the intelligent civilizations of souls. "Soul" is defined here as meaning an individuated, nonlocal, conscious, intelligent entity that is based in the Interlife dimensions and that is a holographic fragment of the original Source or creator of the spiritual dimensions of the Omniverse. The empirical evidence for the existence of souls is derived from a replicable database of more than 7,000 cases of hypnotic regression of soul memories of the Interlife, developed according to a standard protocol. Replicable data report that souls are created in a process that results in a soul as a holographic "egg of Light" drawn from the original Source.[4]

Replicable data from hypnotic regression of soul memories of the Interlife now provide *prima facie* empirical evidence supporting the dimensional ecology of the Omniverse hypothesis, since they provide detailed information regarding the dimensional

---

[4] Newton, Destiny of Souls, pp. 127-128.

INTRODUCTION: A MAP OF THE OMNIVERSE

interactions of the intelligent civilizations of souls, of spiritual beings, and of Source (God) with the universes and intelligent civilizations of the multiverse.

Our map also evaluates the *prima facie* evidence for the multi-dimensional role of souls in the dimensional ecology of the multiverse, finding that the intelligent civilizations of souls in the spiritual dimensions have a central role in the creation and maintenance of the universes of time, space, energy, and matter of the multiverse. The universes of the multiverse serve as a virtual reality within which souls can attain higher degrees of development through the moral experience of incarnations as a diversity of Exophenotype creatures, humanoids and many others.

Souls also participate collectively in the process of creating the universes of the multiverse. There is *prima facie* evidence that the intelligent civilizations of souls are involved in life creation in the universes of the multiverse. Souls are involved in the creation of galactic matter, stars, and planets in the Exopolitical dimensions. Souls undertake interdimensional travel from the Interlife in the dimensional ecology to create or adjust planets in a universe of the multiverse.

Our map of the Omniverse reveals the extent of souls' role in cosmic creation in the multiverse, and why the totality of the spiritual dimensions—souls, spiritual beings, and God—collectively function as the Source of all of the physical universes of time, space, energy, and matter in the Exopolitical dimensions.

Hyperdimensional Civilizations' Role in Dimensional Ecology

Chapter 8 addresses Grey hyperdimensional control over soul reincarnation procedures. Why are some Grey hyperdimensional civilizations "playing God" with human souls? How does the dimensional ecology allow for souls to be based on the spaceships of some hyperdimensional Grey intelligent species? There is *prima* *facie* empirical evidence that human souls are aboard some hyperdimensional Grey spacecraft, undergoing complex soul reincarnation operations that are carried out by the Grey extraterrestrials.

Our map explores a case study of Grey hyperdimensionals implanting the soul of a fetus in the womb of a pregnant human woman aboard a Grey spacecraft. Our map concludes that this case demonstrates that souls and specific Grey hyperdimensionals navigate the dimensional ecology of the Multiverse as interdimensional entities in cooperative relationships, such as mutual assistance in education of a future parent of an incarnating soul and in the actual incarnation process.

The cases of Suzanne Hansen's future son, and of the reported father and grandfather who accompanied a 15-year-old boy during an encounter with Grey hyperdimensionals, as we will see, appear to be part of a structured program of cooperation among the intelligent civilizations of souls and this specific species of Grey hyperdimensionals for specialized intervention in specific types of soul incarnations or educational encounters with Grey hyperdimensionals in the Exopolitical dimensions of the Omniverse.

The Science of the Omniverse

Chapter 9 explores ten top implications for a positive human future that can be drawn from research into the science of the Omniverse and for further exploration of the dimensional ecology of the Omniverse. These ten top implications can be summarized here as follows.

1 A reasonable observer will be able to conclude that *prima facie* empirical evidence supports the dimensional ecology of the Omniverse hypothesis. This hypothesis holds that we earthlings live in a dimensional ecology of intelligent life that encompasses intelligent civilizations based in parallel dimensions and universes in the multiverse as well as souls, spiritual

beings, and Source (God) in the spiritual dimensions. Together, the Exopolitical dimensions and the spiritual dimensions form the Omniverse. The totality of the spiritual dimensions (souls, spiritual beings and God) function as the source of the universes of the multiverse.

2   Exopolitics, the science of relations among intelligent civilizations, and parapsychology, the science of psi consciousness, telepathy, reincarnation, the soul, the Interlife, and Source (God), are among the proper scientific disciplines for exploring and mapping the dimensional ecology of the Omniverse.

3   Dimensionality, the ability of intelligence to organize itself via dimensions (discrete bands of conscious energy), appears to be a key criterion by which the Omniverse is designed, in both the spiritual dimensions and the Exopolitical dimensions.

4   The consciousness and developmental level of souls incarnating on Earth and our collective ability to comprehend the dimensional ecology of the multiverse constitute both a bottleneck and a key to the future evolution of Earthling humans in the society of organized intelligent life in the Exopolitical dimensions.

5   The long-term positive transformation of power, economic, and social structures on Earth may depend on "Soul Power." That is, significant soul development during an incarnation can occur as breakthroughs in how science-based knowledge and information about the dimensional ecology of the Omniverse is consciously internalized and openly acknowledged in human society on Earth.

6   The dimensional ecology of the Omniverse hypothesis and the dimension-based typology of intelligent civilizations in the Exopolitical dimensions can facilitate and accelerate the comprehension of UFO and extraterrestrial-related data and

information by the Earthling human public and by scientific, governmental, educational, and media organizations.

7   The dimensional ecology of the Omniverse hypothesis and the science-based study of the spiritual dimension can facilitate and accelerate comprehension of true versions of basic concepts of reality, such as soul, Interlife and Source (God), by the Earthling human public and by scientific, governmental, educational, and media organizations.

8   The science-based study of the dimensional ecology of the Exopolitical and spiritual dimensions of the Omniverse reveals the centrality of the intelligent civilizations of souls in the creation, maintenance, design, and ultimately incarnation in the universes of the Exopolitical dimensions.

9   Humanity now is being misinformed about the true nature of the soul, of the Interlife, of the mechanisms of reincarnation, and ultimately of Source (God). Religions are a large source of such erroneous information, which is based on texts and ancient religious belief systems that are not scientifically correct and are yet considered sacred, as a matter of faith.

10 The dimensional ecology of the Omniverse hypothesis is itself a matrix for recognition and classification of ongoing and new research into intelligent civilizations in the exopolitical and spiritual dimensions. This hypothesis brings science and spirituality together in a way that will return science to the proper study and understanding of the human soul and, one may hope, restore understanding to supremacy over ignorance. To support and encourage this understanding is our duty as informed, aware human souls.

PART I

THE OMNIVERSE AND ITS INHABITANTS

CHAPTER 1
THE OMNIVERSE

The total number of universes besides our own estimated to exist in the multiverse is staggering. The multiverse is defined as the total of all universes, including our own, and is thought to encompass "all space, time, matter, and energy."[1]

Cosmologists Richard L. Amoroso and Elizabeth A. Rauscher write,

> Generally multiverse, sometimes called meta-universe or megaverse, is the hypothetical set of multiple possible universes (including our Einstein-Hubble universe) that together comprise all of reality...The structure of the multiverse, the nature of each constituent universe, and the connection between them depends on the particular multiverse hypothesis being considered by the theory. But the term universe is supposed to represent the entirety of all existence; however, with usages like "Mr. Tompkins Universe" or the universe of the ant, one has become accustomed to the idea of many universes, at least in the common vernacular. Interestingly, American psychologist William James first coined the term multiverse in 1895. In scientific circles, many disparate definitions of the multiverse exist, such as parallel universes, Bubble universes, alternate realities as in Everett's Many Worlds interpretation of quantum theory containing

---

[1] Clara Markowitz, "Five reasons we may live in a multiverse," http://bit.ly/18Q7t8N.

every possibility, or the ... extension of string theory known as M-theory, where our universe and others are purported to be created by collisions between membranes in an [11-dimensional] space.[2]

NUMBER OF UNIVERSES IN THE MULTIVERSE

Physicists Andrei Linde and Vitaly Vanchurin of Stanford University recently calculated "that the total number of such universes, in the simplest inflationary models, may exceed" a number one can write as

$$10 \text{ raised to the } (10 \text{ raised to } 10^7) \text{ power.}^3$$

This is a deceptively compact notation. First, 10^7 is a 1 with seven zeroes after it, that is, 10,000,000, or ten million. Next, 10 raised to the ten-millionth power, that is, $10^{10,000,000}$, is a 1 with ten million zeroes after it. Written out with six zeroes to the inch, it would stretch for about 26 miles. But the next step, raising 10 to the power of that 26-mile number, generates a number so large that we cannot name it, let alone write it out. It would stretch for at least 260 million miles. Linde and Vanchurin also said, "This humongous number is strongly model dependent and may change when one uses different definitions of what is the boundary of eternal inflation."

One German supercomputer simulation estimates that there are "500 billion galaxies in our universe."[4] Astronomers now estimate there are 100 billion habitable Earth-like planets in our Milky Way galaxy and 50 sextillion habitable Earth-like planets in our particular universe.[5]

---

[2] Amoroso and Rauscher, The Holographic Anthropic Multiverse, p. 32.

[3] Linde and Vanchurin, "How many universes are in the multiverse?" http://bit.ly/HtbdEr.

[4] "500 Billion—A Universe of Galaxies: Some Older than Milky Way," The Daily Galaxy, June 10, 2013, http://bit.ly/1grvKpP.

[5] Anthony, "Astronomers estimate 100 billion habitable Earth-like planets in the Milky Way, 50 sextillion in the universe," http://bit.ly/13UiMgU.

Is Our Universe Finite or Infinite?

Working with the Baryon Oscillation Spectroscopic Survey (BOSS), researchers have now developed an "ultraprecise galaxy map" that has measured the distance to galaxies in our universe more than 6 billion light years away to within 1 percent accuracy. A report on the map stated,

> 'There are not many things in our daily lives that we know to 1 percent accuracy,' David Schlegel, a physicist at Lawrence Berkeley National Laboratory and the principal investigator of BOSS, said in a statement. 'I now know the size of the universe better than I know the size of my house.'
>
> The new results, presented by Schlegel and his colleagues [on January 8, 2014] at the 223rd meeting of the American Astronomical Society, also provide one of the best-ever determinations of the curvature of space, researchers said. In short, the universe appears to be quite "flat," meaning that its shape can be described well by Euclidean geometry, in which straight lines are parallel and the angles in a triangle add up to 180 degrees.
>
> 'One of the reasons we care is that a flat universe has implications for whether the universe is infinite,' Schlegel said. 'That means—while we can't say with certainty that it will never come to an end—it's likely the universe extends forever in space and will go on forever in time. Our results are consistent with an infinite universe.'[6]

This tentative finding, that our universe is infinite, is congruent with Linde and Vanchurin's finding that there are a "humongous" number of universes in the multiverse, since universes that extend forever in space and go on forever in time can coexist in parallel with each other.

---

[6] Miriam Kramer, "Scale of universe measured with 1 percent accuracy," http://bit.ly/1e96Rhq.

Intelligent Civilizations in Our Universe and Multiverse

A conservative estimate of the number of communicating intelligent civilizations in our universe is one hundred billion (100,000,000,000). This estimate is based on the 1960 Drake equation, which assumes that there are only twelve communicating intelligent civilizations in our Milky Way galaxy, out of the estimated 100 billion habitable Earth-like planets.[7] Then how many intelligent civilizations may there be in the multiverse? If we multiply the Drake equation-based estimate of a hundred billion communicating intelligent civilizations in our universe by Linde and Vanchurin's calculation of the number of universes in the multiverse, we arrive at the number of intelligent civilizations in the multiverse as being 100,000,000,000 times that 260-million-mile-long number. It is not physically possible to actually write that number out fully.

Intelligent Civilizations in the Spiritual Dimensions

The universes of the multiverse are not the only dimensions where intelligent civilizations are based. There is *prima facie* replicable empirical evidence of intelligent civilizations that are based in dimensions that are outside the multiverse. We can term these dimensions the "spiritual dimensions."

One important database of such empirical evidence for the existence of intelligent civilizations in the spiritual dimensions is derived from more than 7,000 cases of replicable hypnotic regressions of soul memories of the Interlife (or afterlife), developed according to a standard laboratory protocol by Dr. Michael Newton.[8] This database contains replicable evidence for the

---

[7] Randy Krum, "How many alien civilizations are there in the galaxy?" http://bit.ly/16ygWlH.

[8] See Dr. Newton's books in my source list.

intelligent civilizations of souls, for the intelligent civilizations of spiritual beings; and for the Source (God).

By intelligent civilizations of "souls" I mean civilizations of individuated, nonlocal, conscious, intelligent entities that are based in the dimensions of the Interlife (or afterlife). Each soul, by the evidence, is a holographic fragment of the original Source (God), the creator of the spiritual dimensions.

One hypothesis about the nature of the Source (God) supported by empirical data is based on a replicable finding that the Source (God) originally responsible for the spiritual dimensions manifests as a vast "Sea of Light" within the spiritual dimensions. Souls, including yours and mine, are formed as "eggs of Light" or holographic fragments from that Sea of Light in an as-yet unrevealed process. These replicable findings are that each soul is a holographic fragment of the whole of Source (God).

World Public Opinion and Intelligent Life in the Multiverse

World public opinion is congruent with the recent science-based estimates that there are one hundred billion communicating intelligent civilizations in our universe, and an even more "humongous" number of communicating intelligent civilizations in the multiverse.

The possibility that we live in a populated cosmos is conventionally thought to be controversial and esoteric. Public opinion around the world is divided as to whether we are alone in the cosmos. The world public has been quarantined from real knowledge about the actual role of non-Earth intelligent civilizations on Earth. Instead of public education about an extraterrestrial presence and Earth's history in the galaxy, governments have knowingly fed the world public a steady diet of disinformation and brainwashing about Earth's dealings with intelligent civilizations. Ever since the 1953 US Central Intelligence Agency

Robertson Panel, the facts of intelligent civilizations and their visitations to Earth have been classified and off limits for civil society.[9] Consequently, there is a divided world public opinion about the presence of extraterrestrial and hyperdimensional civilizations in Earth's environment. Nevertheless, there is a core of world public opinion and of public opinion in specific nations which accepts that humanity co-exists in a cosmos populated by other intelligent civilizations.

One 2013 public-opinion poll of 5,886 US adult residents found that "37 percent affirmed a belief in the existence of extraterrestrial life, 21 percent denied such a belief, and 42 percent were uncertain, responding 'I'm not sure'."[10] A 2010 by the French market-research company Ipsos poll of world public opinion on extraterrestrials found that "one in five (20 percent) of presumably human adults surveyed in 22 countries (representing 75 percent of the world's GDP) say they believe that alien beings have come down to Earth and walk amongst us in our communities disguised as 'us'." People in India (45 percent) and China (42 percent) are most likely to believe that extraterrestrials are visiting Earth.[11]

WORLD PUBLIC OPINION AND THE SPIRITUAL DIMENSIONS

According to a 2011 Ipsos poll taken in 23 nations among 18,829 adults, "one half (51 percent) of global citizens definitely believe in a 'divine entity' compared to 18 percent who don't and 17 percent who just aren't sure." The Ipsos poll also found "Similarly, half (51 percent) believe in some kind of afterlife, while the remaining half believe they will either just 'cease to exist' (23 percent) or simply 'don't know' (26 percent) about a hereafter.

---

[9] Report of Scientific Advisory Panel on Unidentified Flying Objects, http://bit.ly/15ry5Na.

[10] Troy Matthew, "Science? Or sacrilege?" http://bit.ly/17omJWM.

[11] Ipsos, "One in five (20 percent) global citizens believe that alien beings have come down To Earth," http://bit.ly/1cx0qWP.

Seven percent of respondents believe in reincarnation."[12] A substantial core of world public opinion thus has a view of reality that is congruent with the *prima facie* replicable empirical evidence for an Interlife (afterlife), intelligent civilizations of souls, civilizations of spiritual beings, and a Source (God) or Creator.

In the United States of America, a December 2013 Harris poll showed that a substantial majority of Americans hold positive beliefs about the existence of God and the essentials of the spiritual dimensions, although the percentage of Americans holding these beliefs has declined somewhat in recent years. The poll finds that 74 percent of U.S. adults do believe in God, but this represents a decline from previous years, when 82 percent expressed a belief in God in 2005, 2007 and 2009. Belief in survival of the soul after death has declined from 69 to 64 percent in recent Harris polls, but 24 percent of Americans believe in reincarnation.

The Harris poll addresses generational and political divides.

> Echo Boomers are less likely than their counterparts in all older generations to express belief in God (64 percent Echo Boomers, 75 percent Gen Xers, 81 percent Baby Boomers, 83 percent Matures)...On the other end of the generational spectrum, Matures are far less likely than any other generation to express belief in ghosts (44 percent Echo Boomers, 46 percent Gen Xers, 46 percent Baby Boomers, 24 percent Matures), witches (27 percent, 29 percent, 28 percent and 18 percent, respectively), and reincarnation (27 percent, 25 percent, 23 percent, and 13 percent, respectively). Turning to the political spectrum, Democrats and Independents show similar levels of belief in most of the tested concepts, with Republicans consistently more likely than either group to express belief in those concepts aligned with the Judeo-Christian belief system; Republicans are less likely than either group to express belief in Darwin's theory of evolution (36 percent Republicans, 52 percent Democrats, 51 percent Independents).[13]

---

[12] Ipsos, "Belief in supreme being(s) and afterlife accepted by half (51 percent) of citizens in 23 country survey," http://bit.ly/19mrjcT.

[13] Harris Interactive Polls, "Americans' belief in God, miracles and heaven declines", Dec. 13, 2013, http://bit.ly/1dne4bJ.

The Harris poll notes that absolute certainty that there is a God is down from what it was 10 years ago.

> In a separate line of questioning, focused on Americans' degree of certainty that there is or is not a God, two-thirds of Americans (68 percent) indicate being either absolutely or somewhat certain that there is a God, while 54 percent specify being absolutely certain; these figures represent drops of 11 and 12 percentage points, respectively, from 2003 testing, where combined certainty was at 79 percent and absolute certainty was at 66 percent. Meanwhile, combined belief that there is no God (16 percent) and uncertainty as to whether or not there is a God (also 16 percent) are both up from 2003 findings (when these levels were 9 percent and 12 percent, respectively). Outside of specific religious samples, the groups most likely to be absolutely certain there is a God include blacks (70 percent), Republicans (65 percent), Matures (62 percent) and Baby Boomers (60 percent), Southerners (61 percent) and Midwesterners (58 percent), and those with a high school education or less (60 percent).

As to the gender of God, the Harris poll notes,

> There continues to be no consensus as to whether God is a man or a woman. Nearly 4 in 10 Americans (39 percent) think He is male, while only 1 percent of U.S. adults believe She is female. However, notable minorities believe God is neither male nor female (31 percent) or both male and female (10 percent). Women, perhaps surprisingly, are more likely than men to believe that God is male (43 percent women, 34 percent men), while men are more likely to believe that God is neither male nor female (34 percent men, 28 percent women). As to God's control over the Earth, there also a continuing—and increasing—lack of consensus as to how much control, if any, God has over what happens on Earth. A 37 percent plurality of Americans (including 52 percent of Catholics) believes that God observes but does not control what happens on Earth, down considerably from 2003, when half of Americans (50 percent) expressed this belief. Just under three in ten (29 percent) Americans, including majorities of those who self-identify as very religious (60 percent) and/or born-again Christians (56 percent), believe that God controls what happens on Earth.

As to religious texts as "Word of God," the Harris poll notes,

> Just under half of Americans believe that all or most of the Old Testament (49 percent) and the New Testament (48 percent) are the "Word of God," representing declines of six percentage points each from 2008 findings. Just under two in ten Americans (19 percent) describe themselves are "very" religious, with an additional four in ten (40 percent) describing themselves as "somewhat" religious (40 percent, down from 49 percent in 2007). Nearly one-fourth of Americans (23 percent) identify themselves as "not at all" religious — a figure that has nearly doubled since 2007, when it was at 12 percent.[14]

Humanity's basic opinions and beliefs about the Interlife (afterlife), soul, and Source (God) are largely formed both by religions and by science. Major religions, such as Christianity, Islam, Hinduism, Buddhism, Judaism, Taoism and Confucianism, Shintoism, and Sikhism, and a dozen medium-sized religions have over 6.5 billion followers, or 90 percent of the total world population of approximately 7.2 billion persons.

Religions are not scientific organizations that reach their conclusions about the Interlife (afterlife), soul, and Source (God) on the basis of laboratory protocols and empirical science. Religions are historical, social, and political organizations whose missions and teaching about the nature of the Interlife (afterlife), soul, and Source (God) are generally based on sacred texts. The content of religions' belief systems about such core issues may differ in detail from religion to religion.

Currently, there is a schism within science as to whether the Interlife (afterlife), soul, and Source (God) actually exist. Academic science largely prohibits research into and teaching about parapsychology and Exopolitics, two sciences necessary for exploring the empirical evidence for such existence. Parapsychology is the science that studies psi consciousness, including telepathy, survival of bodily death, the soul, and reincarnation.

---

[14] Ibid.

Exopolitics is the science that studies relations among intelligent civilizations in the multiverse.

The replicable *prima facie* evidence for the existence of the Interlife (afterlife), soul, and Source (God) is proof that the basic assertions of major religions that these do exist are not based on superstition, ritual, or sociological necessity and are congruent with the discoveries of scientific experimentation.

The Omniverse

The Omniverse is defined as the sum of the multiverse and the spiritual dimensions. The Omniverse contains the totality of all the universes of time, space, energy, and matter, and thus all the intelligent civilizations in the multiverse, *in addition to* the intelligent civilizations of souls and spiritual beings in the Interlife (afterlife) dimensions of the spiritual dimensions, as well as the original Source (God). We can term the multiverse "the Exopolitical dimensions" because the multiverse is the dimensional base for intelligent life in physical form.

Hypothesis: Dimensional Ecology of the Omniverse

Together, the Exopolitical dimensions (the multiverse) and the spiritual dimensions interconnect in the dimensional ecology of the Omniverse. Our human civilization on Earth is in a dimensional ecology that includes the entire Omniverse: our universe, all the universes of the Exopolitical dimensions (the multiverse), and the intelligent civilizations of souls, spiritual beings, and Source (God) in the spiritual dimensions.

An alternative way of stating the hypothesis of the dimensional ecology of the Omniverse is to say that we Earthling humans live within a dimensional ecology in an Omniverse that encompasses:

1. "Exopolitical dimensions," consisting of our known physical universe and multiple parallel universes within which are based extraterrestrial and interdimensional intelligent civilizations, and;

2. "Spiritual dimensions," consisting of dimensions of "afterlifes" or "Interlifes" and other spiritual dimensions within which are based the intelligent civilizations of souls and the intelligent civilizations of spiritual beings as well as a defined Source (God).

A hypothesis is, by accepted definition, "an educated guess, based on observation. Usually a hypothesis can be supported or refuted through experimentation or more observation. A hypothesis can be disproven, but not proven to be true."[15] The dimensional ecology hypothesis is a guide to intelligent life in the Omniverse. It is an educated guess, based on observations such as witness testimony, data, and cases of how, where, and why intelligent life exists in all of the Omniverse.

A simple equation of the dimensional ecology of the Omniverse looks like this.

Omniverse = Exopolitics dimensions (the multiverse)
+ Spiritual dimensions = Dimensional Ecology

The dimensional ecology of the Omniverse hypothesis expands on the conventional scientific definition of the multiverse. Current scientific convention considers the multiverse to consist solely of parallel physical universes of time, space, energy, and matter, of which our physical universe is one. One conventional view of the multiverse is, "The universe we live in may not be the only one out there. In fact, our universe could be just one of an infinite [or

---

[15] See, for example, Helmenstine, "Scientific hypothesis, theory, law definitions, learn the language of science" http://bit.ly/bV90Rj.

finite] number of universes making up a 'multiverse'."[16] This view lists five alternative theories of the multiverse, including the existence of infinite universes, bubble universes, parallel universes ("braneworlds"), daughter universes, and mathematical universes.

The spiritual dimension of the Omniverse provides the energy (energy that human scientists such as Lawrence M. Krauss, author of *A Universe from Nothing*, cannot now account for) needed for the creation and maintenance of each physical universe in the multiverse.

The dimensional ecology of intelligent life in the Omniverse hypothesis can be supported through the empirical scientific method, using the jurisprudential law of evidence that employs witness, documentary, and experimental evidence.[17]

CONVENTIONAL SCIENCE SEEKS A SOLUTION LIKE THE DIMENSIONAL ECOLOGY HYPOTHESIS

While not making specific reference to the hypothesis of the dimensional ecology of the Omniverse, advanced conceptual physicists such as Professor Amit Goswami have argued that contemporary science's assumption that "only matter—consisting of atoms or, ultimately, elementary particles—is real" is inadequate and that a new hypothesis of reality is required.

Professor Goswami writes,

> We cannot connect quantum physics with experimental data without using some schema of interpretation, and interpretation depends on the philosophy we bring to bear on the data. The philosophy that has dominated science for centuries (physical, or material, realism) assumes

---

[16] Clara Markowitz, "Five reasons we may live in a multiverse."

[17] Evidence: "The many types of information presented to a judge or jury designed to convince them of the truth or falsity of key facts. Evidence typically includes testimony of witnesses, documents, photographs, items of damaged property, government records, videos, and laboratory reports. Strict rules limit what can be properly admitted as evidence, but dozens of exceptions often mean that creative lawyers find a way to introduce such testimony or other items into evidence." Cornell University Law School, http://bit.ly/HsRF3f.

that only matter—consisting of atoms or, ultimately, elementary particles—is real; all else are secondary phenomena of matter, just a dance of the constituent atoms. This worldview is called realism because objects are assumed to be real and independent of subjects, us, or of how we observe them. The notion, however, that all things are made of atoms is an unproven assumption; it is not based on any direct evidence for all things. When the new physics confronts us with a situation that seems paradoxical from the perspective of material realism, we tend to overlook the possibility that the paradoxes may be arising because of the falsity of our unproven assumption. (We tend to forget that a long-held assumption does not thereby become a fact, and we often even resent being reminded.) *Many physicists today suspect that something is wrong with material realism but are afraid to rock the boat that has served them so well for so long. They do not realize that their boat is drifting and needs new navigation under a new worldview.*[18] [Emphasis added]

Critiquing Professor Goswami's work, Facundo Bromberg writes,

> Science is experiencing one of the most difficult challenges of all history: to explain humans' most basic capacity, the mental. It was for long thought that human mental abilities were insurmountably difficult to understand and therefore [consigned to the] transcendental and disconnected world and relegated to religion. But the advent of the paradoxes in our understanding of the physical world, plus ... research in artificial intelligence and cognitive science faced science with the necessity of dealing with the mysteries of the mind. Every single discipline in humanity has something to say on the issue: science, religion, art, poetry, etc. But none has yet the capacity of proving its assumptions. If Dyer's AI enterprise is successful in explaining consciousness and the mental, then we scientists, Goswami included, will have to welcome it, but if on the other hand, quantum mechanics appears to be more than a mathematical practical tool and [has] ontological relevance, then scientists, including Dyer, will certainly have to accept it. But science discoveries not only take time but resources, and for that we should try to stop an evidently contradictory theory.[19]

---

[18] Goswami, The Self-Aware Universe, pp. 9-10.

[19] Bromberg. "On Goswami's monistic idealism worldview," p. 11.

The dimensional ecology of the Omniverse hypothesis and the replicable *prima facie* evidence that support it provide the necessary tools for a new hypothesis of reality, as well as for the desired *"new navigation under a new worldview,"* the need for which Professor Goswami has identified.

DIMENSIONAL ECOLOGY: SOURCE (GOD), SOULS, AND THE OMNIVERSE

There now exists replicable empirical *prima facie* evidence that confirms some essential aspects of what major spiritual and religious traditions have taught about the nature of Source (God). This evidence informs us that the Source (God) of the Omniverse consists of the totality of the spiritual dimension. God has empirically been found to comprise the intelligent civilizations of souls, the intelligent civilizations of spiritual beings, and the Source (God) itself. This collective entity of the spiritual dimension has been empirically found to be responsible for the ongoing creation of the physical side of the Omniverse, known as the Exopolitical dimensions.

The totality of the spiritual dimension, including (a) God/Source, (b) the intelligent civilizations of souls, and (c) the intelligent civilizations of advanced spiritual beings, is functionally God/Source and acts collectively for the ongoing creation and maintenance of the Exopolitical dimensions of the Omniverse.[20]

PURPOSES OF THE DIMENSIONAL ECOLOGY OF THE OMNIVERSE

A core mission of the dimensional ecology of the Omniverse appears to be the creation and development of souls and spiritual

---

[20] Newton, op cit.

beings in the spiritual dimensions. The intelligent civilizations of souls and of spiritual beings, along with Source (God), collectively create and maintain the totality of the universes of time, space, matter, and energy in the Exopolitical dimensions (the multiverse).

The purposes of the dimensional ecology of the Omniverse include the facilitation of multidimensional development and moral growth of souls in all dimensions of the Omniverse, through a variety of activities. Souls based in the spiritual dimensions incarnate as intelligent entities in the Exopolitical dimensions, and by acquiring the moral experience of life, for example, as an Earthling human, can advance their individual soul development. The soul is a holographic fragment of Source (God) and, by advancing its development, advances the development of the collective spiritual dimension itself.

THE DIMENSIONAL ECOLOGY AND THE TIME-SPACE HOLOGRAM

The physical universe of time, space, energy, and matter that we Earthling humans inhabit is part of the dimensional ecology of the universes of the Exopolitical dimensions of the Omniverse. The multiverse and our physical universe are holographic.[21]

Our universe includes a time-space hologram, within which our Earth human civilization exists, that is an artificial environment created by a higher intelligence. The higher intelligence that created and maintains our universe, including the time-space hologram that we inhabit, is the spiritual dimension that itself is composed of God/Source, the intelligent civilizations of souls that incarnate in the time-space hologram, and the intelligent civilization of spiritual beings. In "Simulations back up theory that Universe is a hologram: A ten-dimensional theory of gravity makes

---

[21] See Michael Talbot, The Holographic Universe, and Sonia Barrett, The Holographic Canvas.

the same predictions as standard quantum physics in fewer dimensions," Ron Cowen writes, "At a black hole, Albert Einstein's theory of gravity apparently clashes with quantum physics, but that conflict could be solved if the Universe were a holographic projection."[22]

THE TIME-SPACE HOLOGRAM IN OUR UNIVERSE

The time-space hologram we inhabit as Earthling humans is composed of multiple timelines. Time travel and teleportation are methods by which intelligent civilizations navigate the dimensional ecology of both the Exopolitical dimensions (the multiverse) and the spiritual dimensions of the Omniverse. Teleportation consists of point-to-point movement across a single timeline. Time travel consists of movement across more than one timeline.

Teleportation can occur within dimensions, such as the time-space hologram that Earthling humans inhabit, or between dimensions and parallel universes of the Exopolitical dimensions. Teleportation and time travel within and between dimensions of the dimensional ecology can occur through naturally occurring stargates or interdimensional portals within both the Exopolitical dimensions (the multiverse) and the spiritual dimensions.[23]

Replicable empirical studies show that, at bodily death in the time-space hologram, a human soul has the option of teleporting to a dimension within the spiritual dimensions called the "afterlife" or "Interlife" through a natural dimensional portal from the time-space hologram in the Exopolitical dimension to the Interlife in the spiritual dimensions.[24]

---

[22] Ron Cowen, "Simulations," http://bit.ly/1bDPwvE.

[23] See Webre, "Web Bot: Andrew Basiago is predicted 'planetary level' whistleblower for Mars life and time travel," http://exm.nr/1c5NUdW; "Why is Hawking affirming time travel theory and appearing ignorant of DARPA secret time travel?" http://exm.nr/GO2Cvz; and "Second whistleblower emerges to confirm reality of time travel," http://exm.nr/18T86P5.

[24] Newton, op. cit.

Teleportation and time travel in the time-space hologram can also occur with advanced technology. We now know from former chrononaut Andrew D. Basiago, a project whistleblower, that the US Department of Defense DARPA's Project Pegasus had developed teleportation and time travel in the period 1969-72. Mr. Basiago has revealed that between 1969 and 1972, as a child participant in Project Pegasus, he both viewed past and future events through a device known as a "chronovisor" and teleported back and forth across the country in vortal tunnels opened in time-space via Tesla-based teleporters located at the Curtiss-Wright Aeronautical Company facility in Wood Ridge, NJ, and at the Sandia National Laboratory in Sandia, NM.[25]

Basiago reported publicly that he saw a physical copy of my 2005 book, *Exopolitics: Politics, Government and Law in the Universe*, in 1971 in the presence of two other witnesses when DARPA's Project Pegasus used advanced time-travel technology to time-travel teleport the book from at least 2005 back in time to 1971.

As corroboration, during 1971-72, I was General Counsel of New York City's Environmental Protection Administration. At that time, I was invited to give an environmental talk to approximately fifty officials. After 2005, I learned that the fifty officials I had spoken before in 1971-72 were in fact DARPA, US Department of Defense, and CIA officials who had been briefed on my future book *Exopolitics*. This time-travel surveillance intelligence mattered to the US Department of Defense and the CIA. In 1977, while working as a Futurist at the Stanford Research Institute, I became the director of the proposed Jimmy Carter White House Extraterrestrial Communication Study.[26]

Secret U.S. government time-travel technology has generally been used for remote sensing in the time-space hologram. Time

---

[25] Webre, "Time travel and political control," http://bit.ly/LoBF0T.

[26] Webre, "My 1970s meeting with DARPA's Project Pegasus secret time-travel program," http://bit.ly/182i3dE.

travel and teleportation have been weaponized rather than used as technologies for the public benefit, such as public teleportation. Time travel in the form of time loops has been used to sequester military secrets and secret military installations. Time travel has been used for political surveillance and political control of the human population via time-travel pre-identification of future U.S. presidents. Secret US government teleportation has been used to teleport project participants to government project sites. According to Andrew D. Basiago, in the early 1970s, DOD time-travel and teleportation liaison (and Nixon Cabinet member) Donald Rumsfeld decreed that teleportation would be used to deploy troops to the battlefield, and not released to the public despite its potential application as a global non-polluting transportation technology and substitute for the petroleum-driven transportation infrastructure.

The Dimensional Ecologies of Consciousness Literature

Alternative expressions that are congruent with the dimensional ecology of the Omniverse hypothesis can be found in the literature on dimensional ecologies of consciousness.

Dimensional ecologies of consciousness consist of bounded ranges and frequencies of energy. They can contain diverse levels of consciousness and life unique to a dimension and can promote the evolution of consciousness across dimensions. In our universe, dimensional ecologies of consciousness can be constructed around dimensions, such as time and light, that Earthling human science is only beginning to explore. Dimensional ecologies of consciousness in our universe function in effect as dimensional ecologies local to our universe that in turn interface with the larger dimensional ecology of the Exopolitical and spiritual dimensions of the Omniverse.

One function of the local dimensional ecologies of consciousness in our universe is to provide stratification and diversity in the evolution of consciousness within our universe. For example, Earth humans are based in the third time-space dimension of our universe's dimensional ecology of consciousness. More evolved humans, such as Arcturian humans, have described themselves as based in the sixth Light dimension of our universe's dimensional ecology of consciousness, according to one system of consciousness stratification.[27]

In terms of their relative position in our universe's dimensional ecology of consciousness, Arcturian humans are more evolved than Earth humans. Souls choosing to incarnate from the Interlife would experience vastly different lives in the Exopolitical dimensions as Arcturian humans from what they would experience as Earth humans. The soul development for a soul incarnating as an Earth human could in theory be greater in magnitude than that experienced by a soul incarnating as an Arcturian.

Cosmologists Barbara Hand Clow and Gerry Clow have explored a dimensional ecology of consciousness of nine dimensions specific to our universe, galaxy, and our Earth. In describing the dimensional ecology of consciousness in our universe, they state that the first and second dimensions are levels within our three-dimensional Earth (3D), from core to surface; the fourth dimension is the realm of collective thought that bridges the physical and unseen worlds; and the fifth through ninth dimensions are celestial. They suppose that most scientists will be unable to imagine the unseen realms that reach beyond the fourth dimension except by means of mathematics, such as topology and superstring theory.[28]

---

[27] Pereira, Arcturan Star Chronicles.

[28] Clow and Clow, Alchemy of the Nine Dimensions, p. xxiii.

DENSITY-BASED ECOLOGIES OF CONSCIOUSNESS

The literature on the density-based ecology of consciousness system is congruent with the dimensional ecology of the Omniverse hypothesis. The major sources to date of the density-based systems of describing ecologies of consciousness are channeled or interdimensionally derived information. Philosopher George Gurdjieff (1866-1949) also developed systems for stratification of consciousness that are congruent with the dimensional ecology of the Omniverse hypothesis.

Although he does not address the multiple universes of the multiverse or their dimensional ecology with the spiritual dimensions of the Omniverse, cosmologist David Wilcock constructs a density-based model of our universe that is congruent with ecology of consciousness within the Omniverse. Wilcock's model is based on channeled material (*The Law of One*) and on replicable empirical evidence, including that of Dr. Michael Newton. Wilcock writes,

> Although this is not considered a popular notion by any means, there is extensive scientific proof that the universe itself is alive. It is a vast, singular, living being— and we are far more interconnected with it than we may have ever believed. A unified, conscious energy field generates the entire cosmos, and this hidden energy can appropriately be called the Source Field. In this new science, galaxies, stars, and planets are life-forms on a scale we can barely even imagine. The basic laws of quantum physics arrange DNA and biological life out of atoms and molecules we would normally consider to be inanimate.[29]

> Natural stargates, or "time doors,' are also discussed at this point. These natural passageways, which exist throughout the universe and allow us to travel to different times and places, are routinely used by souls as a basic form of travel.[30]

---

[29] Wilcock, The Synchronicity Key, pp. 9-10.

[30] Wilcock, op. cit. p. 185..

The literature on dimensional and density-based ecologies of consciousness can be verified and tested against emerging replicable empirical evidence for the dimensional ecology derived hypnotic regression from soul memories of the Interlife. That literature can also serve as useful guides in suggesting future directions for Interlife (afterlife) research based on soul memories of the Interlife or other quantitative methods.[31]

The literature on density-based ecologies of consciousness describes densities or octaves of levels of energies of consciousness. The ecology of consciousness structured around density is the interaction of levels of consciousness in the universes of time, space, matter, and energy of the Exopolitical dimensions as well as in the spiritual dimensions of the Omniverse.

According to one model of the ecology of consciousness as defined by density, "'density' is more or less the same as 'dimension' …'Density' denotes a vibrational frequency and not a location, which the term 'dimension' implies. The density structure of this reality is primarily expressed in … levels, though each level has sublevels within it. The density scale is a model used to communicate one's perception or orientation in relation to other realities."[32]

In the density-based models of the ecology of consciousness local to our universe, "density denotes a qualitatively distinct level of being. Each density has its own structure of life forms, perception, and typical lessons for the consciousnesses residing in it."[33]

Here we need to present an explicit description of the nature of the various density levels.

---

[31] Newton, op cit.

[32] Lamiroy, Exopaedia, "Density," http://bit.ly/1cF2Z9f

[33] Elkins et al., The RA Materials, Cassiopedia.org, http://bit.ly/GJZMaF

First density. According to density-based models of the ecology of consciousness, the first density is

> the consciousness base of inanimate matter and energy local to our universe in the Exopolitical dimensions. [34]

Second density. According to these models, the second density is

> the base density of the evolution of consciousness in the vegetable and animal kingdoms, for example, local to our Earth and our universe. The evolution of consciousness in the second density concerns biological life, survival, adaptation, competition, group organization as seen with animals, and the like. The soul structure is generally a species soul pool, but as species become more advanced, individual members of the species may differentiate themselves by more varied individual learning.[35]

Third density. According to these models, the third density

> corresponds to living beings that have a degree of individual consciousness and corresponding free will, at least in potential. This implies that a proper being of 3rd density is a moral entity with attendant responsibilities. While the animal does primarily [act] according to the typical behavior of the species, the 3rd density entity is supposed to have reflexive self-awareness and free will and to be thus accountable in terms of karma and soul evolution. [36]

Fourth density. The density-based systems of ecologies of consciousness describe the fourth density as

> a partly physical state where graduates of 3rd density may deepen and perfect their chosen polarity. Service To Self and Service To Other groups are distinct in 4th density and do not automatically come in contact, unless in the context of interacting with 3rd density. [Fourth] density beings enjoy more conscious control over physicality and generally form groups telepathically sharing a common pool of experience while retaining a certain individuality. It seems that 3rd-

---

[34] Elkins et al., op.cit., http://bit.ly/16DYQyB
[35] Elkins et al., op.cit., http://bit.ly/GRF6Oj.
[36] Elkins et al., op.cit.,http://bit.ly/GU5xDv.

density beings may visit 4th-density conditions in the context of UFO abduction, for example. This does not, however, make one a 4th-density being. The density to which one is native depends on development of consciousness.[37]

Fifth density. The fifth density of the density-based system is termed the Interlife (or afterlife) in the dimensional ecology of the Omniverse hypothesis. According to the fifth-density model,

> Souls of 1st thru 4th density find themselves in 5th density between incarnations. This is a contemplation zone where these souls may observe their past/future lives from a purely ethereal state of being. However, for progress to be realized, the souls must incarnate in the density which best corresponds to their level of progress.[38]

Sixth density. According to the density-based system of consciousness, the sixth density is

> a level of non-material existence where souls have outgrown the need to incarnate in any density. Souls of 1st through 4th densities go through incarnations, with a contemplation period as non-material forms in fifth density in between. 6D is the last stage before union with the One, or seventh density. Seventh density would correspond to the one source of all creation. The archangels or solar world of the Gurdjieffian cosmology may refer to sixth density.[39]

Seventh density. According to the density-based system of consciousness, this is

> the level where all is one and one is all, in a practical, real and meaningful sense. There is no longer any difference between thought and reality. This corresponds to a notion of all encompassing god or universe or the Sun Absolute of Gurdjieff.[40]

---

[37] Elkins et al., op.cit.,http://bit.ly/16Eab1Hs.

[38] Elkins et al., op.cit. ,http://bit.ly/1ai22gX.

[39] Elkins et al., op.cit , http://bit.ly/1hLQHd4.

[40] Elkins et al., op.cit., http://bit.ly/GJZMaF.

Exopolitics and Parapsychology as Key Sciences of the Omniverse

Exploring and understanding the dimensional ecology of the Omniverse flows out of correlating parallel research in two key sciences: Exopolitics and parapsychology. Exopolitics is a science that studies relations among intelligent civilizations in the universes of the Exopolitical dimensions of the Omniverse.

Exopolitics as a science

The Exopolitics model was first publicly presented in June 2000 in my lecture at a conference at the University of Wyoming chaired by Professor Leo Sprinkle.[41] The science of Exopolitics formally began with my 1999 book, *Exopolitics: A Decade of Contact*, which was published online in the year 2000 as a free e-book to seed a new paradigm of thought for Earthling humankind.[42] Disclosure of, and governmental and public acknowledgement and understanding of, an extraterrestrial and multidimensional presence on Earth has accelerated since 2000. Exopolitics was nominated for 2005 Word of the Year.[43]

As a science, Exopolitics uses tools familiar to other sciences. These include the philosophy of science, a research database of standard classifications, and the law of evidence as a methodology for evaluating witness and documentary evidence. Exopolitics uses the concepts and structures of cosmology and ontology to frame essential questions.

As a branch of philosophy and science, cosmology addresses the nature of the Omniverse, asking questions about the functions,

---

[41] June 2000 Rocky Mountain Conference on UFO Investigation, University of Wyoming, http://bit.ly/1fM1yHn

[42] Although Universebooks.com took the eBook offline with the 2005 publication of Exopolitics; Politics, Government and Law in the Universe, Part II of Exopolitics: A Decade of Contact is online at http://bit.ly/GVR1dH.

[43] Associated Press, Exopolitics Nominated for 2005 "Word of the Year," http://bit.ly/HaArqd.

components, and purpose of the Omniverse. Ontology as a branch of philosophy addresses issues of being, existence, and reality. As a discipline, ontology grapples with the existence or nonexistence of entities such as intelligent civilizations in the Omniverse and considers the typologies within which such civilizations should be grouped. Exopolitics develops typologies of intelligent civilizations in the dimensions and parallel universes of the physical universes of the multiverse in order to research, map, explain, and understand the dynamics and interrelations of these intelligent civilizations.

The Exopolitics model, like the Dimensional Ecology Hypothesis, is the obverse of the twentieth-century canon of science, which holds that all intelligent life ends at Earth's orbit. The penalty for any professional not toeing this line is probable loss of professional status and possible harassment by government authorities. The universe-centric Exopolitics model holds instead that we live in a highly populated universe in a multiverse, filled with intelligent, evolving civilizations operating under universal laws, having governance systems, and mediated by universe politics.

We on Earth are just becoming aware that we live in a populated universe and multiverse that is part of a dimensional ecology. Like all models, the Exopolitics model provides a path for minds to go where they have not ventured before. The Exopolitics model provides the interface between Earthling human, extraterrestrial, and interdimensional intelligent societies in the universes of the Exopolitical dimensions and in the spiritual dimensions of the Omniverse. As society shifts from a twentieth-century model of canonical science to a twenty-first-century model of science, Exopolitics is the empirically based science that is willing to examine the evidence that the Omniverse is filled with intelligent, evolving civilizations.

The science of Exopolitics also addresses questions raised by empirical evidence and replicable scientific knowledge about intelligent civilizations in the universes of the multiverse, as opposed to personal, metaphysical, religious, or logical knowledge, opinion, or belief. Core Exopolitical questions include: Are there non-Earthling intelligent civilizations in the universes of the Multiverse, and how can we know and test this? What forms do intelligent civilizations take in the Omniverse? How can we develop a typology of intelligent civilizations in the universes of the Multiverse?

Exopolitical research data

Primary databases for Exopolitical research and activity can be derived using the scientific method from a wide variety of sources, including at least the following formal categories of primary research data:[44]

1. Whistleblowers from inside secret extraterrestrial liaison programs
2. Documentary evidence (including photographs, video, radar reports, etc.) from government and official sources
3. Replicable information derived through technology-based, Out of Body experiences, Montauk chair device, Chronovisor, Teleporter, Stargate, Plasma chamber, Jump-room
4. Replicable scientific remote viewing using standard laboratory protocols
5. Voluntary conscious physical contactees
6. Involuntary semiconscious physical contactees (abductees)
7. Voluntary semiconscious alter-physical contactees
8. Voluntary psychic contactees (channelers and visionaries)
9. Neutral psychic contactees
10. Superficial excited witnesses and sightings reports
11. Astute debriefers, debunkers, and interpreters
12. Alien artifacts
13. Independent archeology, including Exoarchaeology

---

[44] See Webre, Exopolitics, p. 92.

14 Shamans and occult societies
15 Science fiction
16 Revelations authorized by universe, galactic, or regional governance authorities.

One key challenge of Exopolitical research is that of evaluating content on cosmological concepts that is derived from the sort of revelations covered by category 16 above. Some sources of interdimensional information seem to provide useful hypothetical models of the organization of the universes of the multiverse, as well as of the dimensional ecology of intelligent civilizations in the multiverse and in the spiritual dimensions of the Omniverse. However, sometimes the interdimensional information appears to be mixed with intentional misinformation or disinformation, the intent of which is not entirely clear, although it may be ultimately intended to protect humans from the consequences of having such information.

A good example of such apparently intentional mixing is the Urantia book, a 2,097-page text of interdimensional information published in 1955.[45] On one level, the Urantia book provides extremely valuable hints about possible cosmological models of the Omniverse. Yet it maintains that the Omniverse is structured around seven Superuniverses encompassing planets of the time-space dimensions. Each Superuniverse is composed of about 100,000 local universes for a total of 700,000 local universes, each of which contains approximately 10 million inhabited or inhabitable planets.[46] This is a far smaller number than the one calculated by Professors Linde and Vanchurin.[47]

Would the interdimensional authors of the Urantia book have engaged in intentional deception about the number of universes

---

[45] Belitsos, "Introducing the Urantia Book," p. 213. The Urantia Book, http://bit.ly/1fCFBIr.

[46] Wyllie, Revolt of the Rebel Angels, p. 4.

[47] Linde and Vanchurin, "How many universes are in the multiverse?" http://bit.ly/HtbdEr.

in the multiverse, and why? Well, one answer is that revelations such as those of the Urantia book are intended to transmit general models of the cosmos as hints to humanity. Perhaps its authors were not permitted to reveal detailed information such as Linde and Vanchurin were to discover. This same pattern has been discovered with Earthling humans who have been contacted by hyperdimensional advanced intelligent civilizations: they have often been given accurate information mixed in with intentional misinformation. One purpose of this strategy may be to protect the contactee from assassination by terrestrial military-intelligence authorities, as will be discussed in more depth in Chapter 5 on intelligent civilizations' governance authorities.

Parapsychology as a science

Parapsychology employs the empirical scientific method to study Omniverse phenomena such as psychic capacity, reincarnation of the soul, life after death, the Interlife, the human soul, spiritual beings, the spiritual dimensions of the Omniverse, and God or Source. Parapsychological researchers include J.B. Rhine (psychic capacity), Chris Carter, Ian Stevenson (reincarnation), Raymond A. Moody (near-death experiences), and Michael Newton (the Interlife).

The findings from these two key sciences, Exopolitics and parapsychology, are necessary to address core questions of the dimensional ecology of the Omniverse, such as:

- How do intelligent civilizations in the Exopolitical and spiritual dimensions of the Omniverse interact with each other?
- What is the relationship between intelligent extraterrestrial or interdimensional civilizations and the reincarnation of the soul?
- How do intelligent civilizations in the spiritual dimensions influence intelligent civilizations in the Exopolitical dimensions and *vice versa*?

Core cosmological questions for parapsychology include:

- How can we know and test whether the Omniverse includes a dimension of the Interlife, or life after bodily death?
- How does the typology of intelligent life in the Interlife dimension—the life of the Soul—relate to a typology of intelligent life in the physical dimensions of the Omniverse—the life of the 'Body'?
- What does an integrated model of science-based knowable, provable intelligent life in the Omniverse look like?

The scientific method, the law of evidence, and empirical findings from the emerging sciences of Exopolitics and parapsychology all support the hypothesis that intelligent civilizations interact in a dimensional ecology in the Omniverse.

Overview: Dimensional Ecology of the Omniverse

There is a dimensional ecology among the intelligent civilizations of the universes of the multiverse (the Exopolitical dimensions) and the intelligent civilizations of souls, of spiritual beings and of Source (God) in the spiritual dimensions. The Omniverse contains the totality of all the universes of time, space, energy, and matter and intelligent civilizations in the multiverse *in addition to* the intelligent civilizations of souls, in the Interlife (afterlife) dimensions, and of spiritual beings and Source (God) in the spiritual dimensions.

The dimensional ecology of the Omniverse hypothesis is based on replicable evidence developed according to the empirical scientific method and the law of evidence about subject matters that traditionally have been issues of personal or institutional belief and religious dogma, particularly issues that relate to the existence of the soul, the Interlife (afterlife), and Source (God).

Replicable empirical *prima facie* evidence now confirms that we Earth humans possess souls. The concept that our souls are created as holographic fragments of the original Creator, known in many cultures as God (or variations on that name) is a hypothesis that is supported by replicable empirical experimentation, such as that of Michael Newton. Such evidence now confirms that our souls teleport through interdimensional portals between the Interlife (afterlife) dimensions and our lives as earthly humans and as extraterrestrial and interdimensional beings in parallel dimensions and universes that we call the "Exopolitics" dimensions.

Awareness of the essential elements, if not the details, of the dimensional ecology of the Omniverse has existed in diverse and fragmentary forms for millennia in our earthly religions, on our sacred and occult texts, and in the group and individual beliefs of human societies throughout history. The dimensional ecology of the Omniverse is thus both a scientific hypothesis and a truth that many Earthling humans are familiar with intuitively or have learned during their lives in their families, schools, religions, reading, or philosophizing.

Humans are familiar with many concepts about the dimensional ecology, such as afterlife, souls, and Source (God), although they have widely differing opinions, depending on race, nationality, gender, and education, about whether such concepts actually exist and in what specific form. Many assertions about specific aspects of the dimensional ecology, such as life after death, are set out in the religious beliefs and sacred texts that organized religions have relied on over millennia to win followers. However, important details about, for example, the nature of the afterlife as described by various religions are sometimes factually incorrect when compared to replicable, scientifically derived knowledge about what happens after physical death. In what follows, we will be comparing traditional beliefs with what we can now know from scientific research.

CHAPTER 2
TYPOLOGY OF INTELLIGENT CIVILIZATIONS

In August 2009, an editor for Oxford University Press asked me to write a book applying the principles of international law to extraterrestrial contact. I discovered that Oxford University Press wanted a book exploring how the principles of Earth's international law might apply within our universe as well as in all the universes in the Exopolitical dimensions of the Omniverse.

The editor emphasized to me that

> I have known about your work for some time, and I would like to know if you have the time to do [a book] to encompass an application of the principles of international law to extraterrestrial contact. This work would be a scholarly monograph written to draft suggestions for a legal and diplomatic framework between earth representatives from the UN/various governments, and extraterrestrial intelligences. ... (I personally believe the evidence for their existence is beyond a reasonable doubt ... Without getting into a debate on whether such intelligences exist, ... the book would prepare policy makers, lawyers, law teachers, and diplomats with the necessary background to navigate such contacts.[1]

---

[1] Correspondence from Oxford University Press to author, August 2009.

At the time, it appeared that Oxford University Press was preparing its strategic plan to include science-based and policy works for the post-extraterrestrial disclosure environment. Given that I was a founder of the scientific discipline of Exopolitics, it made sense that Oxford would approach me.

As conceptual work began for such a legal and diplomatic framework between earth representatives from /various governments, and extraterrestrial intelligences, it appeared that no adequate typologies of extraterrestrial intelligences had so far been developed for this task. A new typology of extraterrestrial intelligences needed to be created and developed.

Conventional typology is "the study of types." In anthropology a typology is "division of culture by races," in archaeology a typology is "the classification of things according to their characteristics," and in psychology a typology is "the classification of people according to personality types."[2]

Existing typologies of intelligent civilizations are limited and do not reflect the manner in which intelligent civilizations exist and are self-governed in our universe. The intelligent civilizations in our universe self-report themselves as being based in a specific dimension or density in the in the dimensional ecology of consciousness. That specific dimension or density is that civilization's most fundamental typological, locational, and developmental criterion.

Accordingly, I (with immense help from colleagues) have developed a dimension-based typology of intelligent civilizations, using recent whistleblower, direct witness, and documentary evidence as sources. This new typology establishes dimension-based criteria for extraterrestrial and interdimensional civilizations and extraterrestrial governance bodies.[3]

---

[2] Typology. http://bit.ly/1bbmDDI.

[3] See Webre, "Exopolitics researcher develops evidence-based typology of extraterrestrial civilizations," http://exm.nr/1fy4bwn

Kardashev Scale

An earlier, complementary typology for mapping intelligent civilizations is the Kardashev scale typology. The Kardashev scale was developed by Nikolai Kardashev and popularized by Dr. Michio Kaku. It is based on "a civilization's level of technological advancement," rather than on the more fundamental criterion of dimension or density in the dimensional ecology of consciousness in which the civilization is based. The Kardashev scale is a method of measuring an advanced civilization's level of technological advancement. The scale is only theoretical and in terms of an actual civilization highly speculative; however, it puts energy consumption of an entire civilization in a cosmic perspective. It was first proposed in 1964 by the Soviet Russian astronomer Nikolai Kardashev. The scale has three designated categories called Type I, II, and III. These are based on the amount of usable energy a civilization has at its disposal, and the degree of space colonization. In general terms, a Type I civilization has achieved mastery of the resources of its home planet, Type II of its solar system, and Type III of its galaxy.[4]

Cooperating with the Military-Industrial-Extraterrestrial Complex

Another earlier, complementary typology classifies intelligent civilizations by race or approximate location in the universe. This typology also asks whether a specific extraterrestrial civilization is "cooperating with or outside Earth's Military-Industrial-Extraterrestrial Complex." [5] This Earth-centric typology does not include the dimensional anchoring of intelligent civilizations as a fundamental criterion of a

---

[4] Kardashev 1964. Garland, "The Kardashev Scale," http://bit.ly/1bNRJBm. Dvorsky, "How to measure the power of alien civilizations using the Kardashev scale," http://bit.ly/J3G0Yy.

[5] Salla, "A report on the motivations and activities of extraterrestrial races," http://bit.ly/H0ajyp

civilization's reality. The typology is limited to those civilizations that are in relationship to the governing complex of one nation (USA) on one planet (Earth) in one solar system (Sol) in a galaxy (Milky Way) that has been shown to have at least 500 million Earth-like planets and a universe that may have hundreds of billions of galaxies. This specific typology is too Earth-centric for application to our universe, let alone the Omniverse.

DIMENSION-BASED TYPOLOGY OF INTELLIGENT CIVILIZATIONS

The dimension-based typology of intelligent civilizations establishes an evidence-based typological model for intelligent civilizations and for intelligent civilizations' governance authorities. This new typology of intelligent civilizations is based on how advanced civilizations describe themselves. The Exopolitical literature shows, for example, that advanced intelligent civilizations describe themselves by the dimension in the dimensional ecology of consciousness in which they are based. For example, "I am a Pleiadian of the fifth dimension" or "I am an Arcturian of the sixth dimension." Our Earth's time-space dimension has been variously described as the third or fourth dimension in the dimensional ecology of consciousness.[6]

The new dimension-based typology of intelligent civilizations integrates whistleblower, direct eyewitness, and documentary evidence for intelligent civilizations in the parallel universes of the multiverse. The new typology establishes an evidence-based typological model for intelligent civilizations and for intelligent civilizations' governance authorities.

---

[6] See, for example, Clow and Clow, Alchemy of Nine Dimension and The Pleiadian Agenda; Brown, Cosmic Voyage and Cosmic Explorers; and Pereira, Arcturan Star Chronicles.

TYPOLOGY OF INTELLIGENT CIVILIZATIONS

In outline format, the dimension-based typology of intelligent civilizations in the Exopolitical dimensions of the Omniverse is as follows:

A   Intelligent civilizations (time-space dimension)
  1   Solar-system civilizations based in the time-space dimension, such as
    a   our human civilization here on Earth, and
    b   the intelligent human civilization living under the surface of Mars that reportedly enjoys a strategic relationship with the United States government.
  2   Deep-space civilizations based in the time-space dimension on a planet, solar system, or space station in our or another galaxy or in some other location in this known physical universe.
B   Hyperdimensional civilizations: intelligent civilizations that are based in dimensions higher that our time-space dimension or in other parallel universes and that may use technologically advanced transport or physical forms when entering our known time-space dimension.
C   Intelligent civilizations' governance authorities.

For intelligent civilizations' law and governance, the dimension-based typology identifies legally constituted governance authorities with jurisdiction over a defined dimensional component of our universe or another parallel universe of the multiverse. Replicable, empirical communication with representatives of governance authorities of our Milky Way galaxy has been reportedly achieved using standard laboratory remote-viewing protocols and has been reported in the public domain[7]

As verifiable disclosures of classified US and other governmental reports on human-extraterrestrial liaison and planetary

---

[7]   Brown, Cosmic Explorers, pp. 147-177.

visitation programs are made public, directly or through whistleblower eyewitnesses, the dimension-based typology can be expanded and refined. Likewise, as more data about the nature of intelligent civilizations in universes parallel to our own are gathered, the dimension-based typology of intelligent civilizations in the universes of the Multiverse can be refined and expanded accordingly.

EXOPHENOTYPOLOGY: INTELLIGENT CIVILIZATIONS IN THE DIMENSIONAL ECOLOGY OF THE EXOPOLITICAL DIMENSIONS

The dimension-based typology of intelligent civilizations in the multiverse helps us identify the dimension, density, or strata of the local dimensional ecology of consciousness in our universe (or other universes) in the Exopolitical dimensions in which a particular intelligent species of civilizations is based. Concurrent with its dimension, an intelligent species or civilization can also be typed by Exophenotypology, which is the typology or classification of extraterrestrials based on their physical appearance that has been developed by Dr. Manuel S. Lamiroy of Exopolitics South Africa.[8] According to Lamiroy, Patrick Huyghe and Dennis Stacy were probably the first to make a systematic classification based on phenotypes, i.e., the observable characteristics of aliens.[9]

Higher intelligence and intelligent civilizations can manifest in our universe and universes of the multiverse in many different exophrenotypes. A preliminary exophenotypology of intelligent civilizations in the Multiverse has been developed, based on observed intelligent civilizations species.[10]

---

[8] Lamiroy, "Exophenotypology," http://bit.ly/1giGxFn.
[9] Huyghe and Stacy, Dickinson and Schaller, as cited by Lamiroy, http://bit.ly/1giGxFn.
[10] Ibid.

Humanoid exophenotype

According to exophenotypological observation, humanoid exophenotypes are those who physically resemble our species of human beings. Four subcategories of the humanoid exophenotype can be distinguished, as follows.

Standard terran humanoid. These are the extraterrestrials who look identical to terrestrial human beings and who can therefore walk among us unrecognized. This subcategory includes the so-called Blond or Nordic Extraterrestrials who are often observed; as well as many others.

A humanoid exophenotype observed on Earth is the Sasquatch, Ogo, or Yeti. These are three names for what is likely just one species. We seem to be dealing with the same kind of creature but in three different locations and with a different color of fur depending on the location: white fur in the Himalayas, reddish brown to black fur in North America, and black fur in Africa. People also refer to these beings as "Bigfoot" and, inspired by *Star Wars*, as "Chewbaccas" or "wookies."[11]

Shorter humanoid. Terrestrial-looking extraterrestrials who are shorter than most terrestrial humans. These include, e.g., the Santinian, the Korendrian, and the Short Non-grey.

Taller humanoid. This subcategory consists of extraterrestrials that look terrestrial in appearance, but are taller than the average human being. These include the Tall White, as well as several types of "giants" (e.g., the Bawwi).

Humanoid with different features. A fourth subcategory consists of beings that in many ways look Terrestrial humans, yet have one or more clearly distinguishing features, such as colors or types of skin, hair, or eyes that are not found here on Earth. These include, e.g., cat people, bird-people, blue-skinned aliens, and aliens with wrinkled faces.[12]

---

[11] Lamiroy, "Non-humanoid mammalian phenotype," http://bit.ly/16RSNXn.

[12] Lamiroy, "Humanoid phenotype," http://bit.ly/19PpukK.

Three Martian humanoid exophenotypes, discovered by former US chrononaut Andrew D. Basiago and discussed at length in the next chapter, also fall under the humanoid exophenotypes:

- Homo martis terris
- Homo martis martis
- Homo martis extraterrestrialis

Grey hyperdimensional exophenotype

According to the Exophenotypology developed by Lamiroy, "The Greys are probably the best-known phenotype of extraterrestrials [or hyperdimensionals]. Older UFO literature distinguishes three main types, but closer inspection reveals there are actually five, often with many variations between them, especially for the "standard" Greys, who differ in skin color, in whether their arms seem to have an elbow or not, in whether they have hands or claws, and in the appendages on those hands, which can be fingers or tentacles with suction caps. The five main types are as follows:

Standard Greys are the most common type. They are about 4 to 4.5 feet tall, with large bulbous heads and wraparound eyes, a slit-like mouth, small ears without earlobes, and no visible nose. Their legs are shorter and jointed differently than one would expect in a human. Their arms often reach down to their knees. These Greys have been seen in many different colors and are generally pale in appearance: grey, white, (pale) blue, (pale) green, (pale) orange, and brown. There does not seem to be one standard phenotype of these, as they have been observed with many different features: some have hands with fingers, some have claws, some have webs, etc.

Tall Nosed Greys are 6 to 9 feet tall and, unlike other Greys, have a distinct nose.

Tall Greys are 6 to 7 feet tall, and basically are taller versions of the Standard Greys. They have large heads, wraparound eyes, and a slit-like mouth. We may here be dealing with two different types, because, unlike the Standard Greys, some of these Tall Greys are said to have reproductive organs. They also seem to have something that resembles a small nose.

Short Greys are shorter, about 3.5 feet tall. They typically look like shorter, but far more muscular versions of the Standard Greys. They are said to be extremely aggressive and generally are believed to be the most dangerous of all grey species. (On the Internet, these are often dubbed "Bellatrix Greys.")

Mini Greys: in recent years, there have been reports of Greys that fit the description of the Standard Greys, but are only two feet tall instead of the usual 4 to 4.5.[13]

Although some entities known as Grey hyperdimensionals may be artificial cyborgs or probes, the Grey races, which one expert numbers in excess of 150, are living species. A dialogue with a Short Grey sponsored by the Exobiology Project in 2006 suggested that Greys, while not plants, have a biology that thrives on chlorophyll, as this brief excerpt shows.

> Q. Are you a mechanical species, as some Humans claim?
>
> A. *Small Grey*: No, we are biological but different in our biology than Human and other species.
>
> Q. Are you part plant, as some humans say, that your tissues contain chlorophyll?
>
> A. *Small Grey*: We are not plant but use chlorophyll in a symbiotic way. Chlorophyll provides us with energy and nutrients.
>
> Q. Is that why you look bluish-grey in white light?
>
> A. *Small Grey*: Yes, the chlorophyll in our skins colors our appearance. We are also able to bioluminescence.
>
> Q. I once "observed" a Grey in captivity. Did this one die and what could have been done to keep it alive?

---

[13] Lamiroy, "Grey phenotype," http://bit.ly/19PpukK and http://bit.ly/H6Dksh.

A. *Small Grey*: It did die. It needed other Greys and it needed nutrients.

Q. Do we have these nutrients on Earth?

A. *Small Grey*: Some of them. Chlorophyll, boron, but others you do not have.

Q. So, that means that captured Greys will always die?

A. *Small Grey*: Not always. Some may have enough reserves to survive.

Q. Are you mainly nocturnal?

A. *Small Grey*: Yes, we work by night. Bright light, unless it has special filters, damages us. Both heat and light damage us.

Q. How does the chlorophyll provide nutrients if it does not have sunlight?

A. *Small Grey*: We have special illumination."[14]

Reptilian exophenotype

Current literature about alien abduction and encounters mentions beings that are reptilian in appearance, being called reptilian, reptoid, dinoid, or saurian. Virtually all of those mentioned below are bipedal—they walk upright—and therefore are not really reptiles, but instead have characteristics reminiscent of reptilians. The main two exceptions to this would be the Ciakar and the "giant snakes," both of whom fit the description of actual reptiles.

According to the Exophenotypology developed by Lamiroy, the following subcategories of the reptilian exophenotype can be distinguished.

Tall Winged Draco (Ciakar). These have been mentioned in the literature, but have not actually been encountered on Earth. They look like giant lizards, range from 14 to 22 feet tall, can weigh up to 1,800 pounds, and can have winged appendages.

---

[14] Smith, "The Exobiology Project: Talking with the Small Greys."

Winged Draco / Mothmen. Reports about flying reptilians with wings originate mainly in the Americas and in Asia. These reptilians are bipedal and have a very muscular, athletic build. They stand 7 to 8 feet tall. Their heads are typically said to be dinosaur-like (like Tyrannosaurus Rex). One feature mentioned again and again is that they have glowing red eyes with a hypnotic stare. Like small Greys, they are said to be able to fly or levitate. They have wings on their backs, but thus far there have been no reports of these wings moving or being used, not even when the beings are flying. The literature often refers to these beings as Mothmen.

Non-winged Draco. The non-winged Draco look much like the winged Draco but lack wings and have different colors of eyes. The non-winged Draco are one of the reptilian species most often seen on board ships that abductees are taken to. As such, they are often seen in company of Greys. Reports of these non-winged Draco come from all over the world. They are among the most commonly encountered reptilians. The reports are contradictory, however, over whether these reptilians have genitalia. Some accounts mention no genitalia, while others do mention their having genitalia and engaging in sexual acts with abductees.

Iguana-like. The literature also mentions reptilians who are described as iguana-like. They look similar to the non-winged Draco, but are less muscular. Their heads resemble that of an iguana more than that of a Tyrannosaurus. They are also shorter than the typical non-winged Draco, being about 5 feet tall. In most reports, they are said to be hooded.

Lizard-like. A fifth subcategory of reptilians consists of lizard men / snake men. They are about 6 to 8 feet tall, upright, bipedal, often have lizard-like scales, are greenish to brownish in color, and have claw-like, four-fingered, webbed hands. Their faces are said to be a cross between that of a human and that of a snake or lizard. Some are said to have a central ridge coming down from the top of the head to the snout. Since some witnesses refer to a "snake-like" head, whereas others use "lizard-like," these may be two separate groups. Given the different descriptions of skin and genitalia, we may actually be dealing with several different groups.

Giant snakes. Some unconfirmed reports on the Internet mention gigantic snakes of extraterrestrial origin. These would be genuine reptiles, not bipedal, not even legged.

Others. Abductees have also described encounters with reptilian-looking beings who do not fall in the categories listed above. Some of them seem to be variations on the other phenotypes. E.g., there are accounts of beings who fit the description of the lizard men, but who seem to have a wet leathery skin and hands with fingers instead of claws.[15]

A 2006 dialogue conducted by the Exobiology project with a reptilian hyperdimensional set out several possible characteristics of this species.

> Q. You sound mechanical, according to Earth characteristics. Are you mechanical or physical?
>
> A. *Hooded Reptilian*: We are physical. We are not mechanical.
>
> Q. Why are some species interested in hydrocarbons? Are they rare in the galaxy? What about other minerals and gases: hydrogen, boron? What purpose do these species put hydrocarbons to? Are we in conflict for these resources?
>
> A. *Hooded Reptilian*: The Greys bring Earth matters to us for fuel, for nurture, for cleansing, for breathing, for sustenance, for "greening."
>
> Q. What is "greening"?
>
> A. *Hooded Reptilian*: We have chlorophyll in our skins for nutrition.
>
> Q. Like the Greys?
>
> A. *Hooded Reptilian*: Yes. There is no conflict for resources. Resources are everywhere.
>
> Q. That is what the other races have said.
>
> A. *Hooded Reptilian*: That is true.[16]

---

[15] Lamiroy, "Reptilian phenotype," http://bit.ly/1czbS1F.

[16] Smith, "The Exobiology Project: Talking with the Hooded Reptilian."

The Lamiroy Exophenotypology also develops Insectoid and Hybrid typologies.

Insectoid exophenotype

The insectoid phenotype consists of those beings whose appearances (and/or behavior) resemble those of insects. As with the reptilians and amphibians, the insectoids are a mixed category, in which some phenotypes display some humanoid traits, whereas others are more akin to what is normally considered typical for an insect.

Several types have thus far been included:

Praying mantis-like. The most common insectoids are beings who are reminiscent of praying mantises. They are encountered quite often during abduction experiences. Most accounts describe them as approximately 6 feet tall. Most reports list them as bipedal, though some others describe them as more insect-like. It is possible the bipedal ones could be hybrids.

Depanoid. There are reports (and photographs) of beings who are bipedal, with a humanoid torso and limbs, but with an oversized "bug" head. These, too, have been encountered by a handful of abductees. The name "Depanoid" was given by one of those abductees, Jim R. The beings are described as "energy-suckers" or "energy vampires," who put their victims through experiences that the victims respond highly emotionally to; then the beings "feed" off those emotions."[17]

Hybrid Exophenotype

A hybrid is anything derived from heterogeneous sources or composed of elements of different or incongruous kinds. The hybrids spoken of in UFO literature are primarily a cross between

---

[17] Lamiroy, "Insectoid phenotype," http://bit.ly/19ZhN1u.

Earth human and Zeta Reticuli beings. The specific process that is used to create these hybrids has not yet been revealed. According to Dr. Lamiroy, it uses not only genetic splicing and cloning, but a form of light-plasma engineering technology with which humans are unfamiliar.[18]

There also exist exophenotypes for cetaceans (whales and dolphins), amphibians, robots, clones, orbs, shadow beings, stick figures, and nonphysical beings and groups.

EARTHLING HUMAN SIGHTINGS OF AND CONTACT WITH VARIED EXOPHENOTYPES OF INTELLIGENT CIVILIZATIONS

As will be discussed more fully in Chapter 4, there is a broad literature documenting Earthling human sightings of and contact with a variety of Exophenotypes of intelligent civilizations.[19] A standard research database on sightings of possible intelligent civilizations from the dimensional ecology states "The number of documented general sightings in the modern age numbers approximately 120,000, of which 20,000 have been described as landings. Numerous surveys and opinion polls conducted over the last 50 years consistently report that between [5 to 10 percent] of the US population has seen what they consider to be a UFO. According to the United Nations, since 1947, approximately 150 million people have been witnesses to UFO sightings throughout the world."[20]

About Earthling human sightings of other humanoid Exophenotypes, the research data base states, "There are at least 4,000 humanoid cases (researched by Albert Rosales), also known as Close Encounters of the Third Kind (CE-III), a term coined by J. Allen Hynek. 'Humanoid' means 'having human form or characteristics.' The term 'Close Encounter of the Third Kind'

---

[18] Lamiroy, "Hybrid phenotype," http://bit.ly/17VJEc5.

[19] See, for example, Huyghe, Field Guide to Extraterrestrials, and Dickinson and Schaller, Extraterrestrials: A Field Guide for Earthlings.

[20] UFO Evidence, http://bit.ly/18RXnFH.

means seeing or coming in contact with an entity of unknown origin. The encounter may or may not be related to the sighting of an unidentified craft."[21] For sightings by qualified pilots, the research data base states, "There have been over 3,500 documented sightings of Unidentified Aerial Phenomena by military, civilian, and commercial airline pilots. These observations span the entire history of powered flight. Many of these cases come from declassified US government reports and investigations, international reports from official sources, and the direct testimony of military and commercial pilots, air traffic controllers, and radar operators."[22]

On Earthling human contacts with a diversity of Exophenotypes, Lamiroy recently wrote, "Work on the typology has been partially suspended since March 2012 after coming across a database of more than 17,000 contact cases, which literally includes dozens of phenotypes that will still have to be integrated in the exophenotypology."[23]

From the platforms of the dimensions in which they are based, various hyperdimensional civilizations reportedly carry out "interdimensional teleportation" on large numbers and categories of Earthling humans.[24] There is replicable *prima facie* empirical evidence to show that a low estimate of 2 percent of the U.S. population and a high estimate of 14.7 percent of the total human population may have had their physical bodies, etheric bodies, or consciousness temporarily or permanently teleported to spaceships or dimensions other than our time-space Earth dimension by hyperdimensional civilizations.[25] The eyewitnesses

---

[21] UFO Evidence, http://bit.ly/19Ogyxj.

[22] UFO Evidence, http://bit.ly/18MNWV1.

[23] Lamiroy, "Exophenotypes," http://bit.ly/1giGxFn.

[24] Webre, "Up to 1 billion humans are abducted by hyperdimensional ETs," http://exm.nr/HfEpOC.

[25] Webre, "Report: Between 2% of US population and 14.7% of global population are being involuntarily teleported (abducted)," http://bit.ly/1f4GcCi.

to these teleportations reply that they are carried out by various species of intelligent hyperdimensional civilizations of the Grey Exophenotype, as well as by hyperdimensional civilizations of other Exophenotypes. Expert estimates of the number of known Grey intelligent species range as high as 150 different species.[26]

---

[26] Webre, "Mary Rodwell," http://bit.ly/17tBY7B.

PART II

Exopolitical Dimensions of the Omniverse

Chapter 3
Time-Space Solar-System Civilizations

Case studies of intelligent civilizations based in the time-space dimension of a solar system support the dimensional ecology hypothesis for the multiverse. This case study explores the *prima facie* evidence for an intelligent civilization based in the time-space third dimension in our solar system.

In later chapters, we will also explore the *prima facie* evidence for hyperdimensional civilizations and for their interacting via a dimensional ecology with time-space intelligent civilizations in our universe in the Multiverse.

Case Study: Humanoid Exophenotypes on Mars

In our solar system Sol, there is newly emerging *prima facie* empirical evidence that time-space- dimension intelligent indigenous human and humanoid civilizations exist on Earth's nearest planetary neighbor, Mars. This conclusion is the result of empirical observation and application of the law of evidence, and strongly supports the dimensional ecology of the Omniverse hypothesis.

New scientific data and the law of evidence, when applied to recent *prima facie* witness and documentary evidence, establish that an indigenous human civilization, as well as indigenous, intelligent extraterrestrial life, exists on Mars, based in cities beneath the surface of Mars, along with humanoid species on the surface of the fragile ecology of Mars.[1]

This new generation of data from Mars supporting the existence of Martian life has been emerging, first, from the NASA Mars rovers Spirit, Opportunity, and Curiosity that have been on the Martian surface, photographing it at close range, and, second, from independent whistleblowers who were involved in compartmentalized secret US government Mars programs. Application of the law of witness and documentary evidence is a methodology leading to breakthroughs in evaluating this new generation of data.

Proof of the Existence of intelligent Human Martian life

Just as skeptics state that the available evidence proves there is no current indigenous intelligent life on Mars (other than the secret US space program), one can present forensic evidence for the opposite of this assertion and thus prove via *prima facie* evidence that there is indigenous intelligent life of the humanoid Exophenotype currently on Mars. All sides of the debate about the existence of such life deserve to be respectfully heard according to the methodology of the law of evidence.

Exopolitics and the Law of Evidence

Exopolitics employs the law of witness and documentary evidence as a methodology for evaluating this new generation of empirical data about life on planets neighboring Earth and in dimensions

---

[1] Webre, "New data, law of evidence support view of Mars having indigenous, intelligent extraterrestrial life," http://bit.ly/P9nxw6.

adjacent to Earth. In law, *"prima facie* evidence" means that enough evidence, on the face of it, has been presented to support a claim.[2]

Witness evidence

One source summarizes the principles of the law of witness evidence as follows:

> In systems of proof based on the English common law tradition, almost all evidence must be sponsored by a witness, who has sworn or solemnly affirmed to tell the truth. The bulk of the law of evidence regulates the types of evidence that may be sought from witnesses and the manner in which the interrogation of witnesses is conducted during direct examination and cross-examination of witnesses. Other types of evidentiary rules specify the standards of persuasion (e.g., proof beyond a reasonable doubt) that a trier of fact such as a jury must apply when it assesses evidence.[3]

Documentary evidence

Documentary evidence, including photographic evidence, is defined as:

> Evidence of an indirect nature which implies the existence of the main fact in question but does not in itself prove it. That is, the existence of the main fact is deduced from the indirect or circumstantial evidence by a process of probable reasoning. [A] defendant's fingerprints or DNA sample [are] examples of circumstantial evidence. The fact that a defendant had a motive to commit a crime is circumstantial evidence. It is possible to argue that all evidence is ultimately circumstantial, on the premises that no experience whatsoever can directly prove a fact. Recall, however, that courts of law deal with what is reasonable, not with ontology.[4] ... The principal questions of ontology are "What can be said to exist?" and "Into what categories, if any, can we sort existing things?"[5]

---

[2] Legal Dictionary, "Prima Facie," http://bit.ly/1cKCWuW
[3] Law of Evidence. http://bit.ly/1cPKqNg
[4] Ahaze Wetunde, "What are rules governing the admissibility of evidence in court?" http://bit.ly/1cRN8Tk.
[5] Ontology, http://bit.ly/16nUiyh.

At this historical stage in the exploration of Mars, the appropriate standard of proof in determining whether the available witness and documentary evidence shows that indigenous intelligent life does exist on Mars is *prima facie* evidence. There is no scientific, legal, or ethical requirement that such a determination must satisfy the whimsical criterion proposed by Carl Sagan that "Extraordinary claims require extraordinary evidence."[6]

Solar-System Civilizations (Time-Space Third Dimension)

Our initial focus in the dimension-based typology of intelligent civilizations in the Omniverse is the solar-system civilizations based in the time-space dimension in our solar system. These include two intelligent time-space dimension civilizations for which we have empirical evidence, the first being our human civilization here on Earth, the second being the intelligent human civilization living under the surface of Mars that reportedly enjoys a strategic relationship with the United States government. One can demonstrate the existence of our human civilization on Earth from self-evident common sense, empirical science, and the law of evidence. For example, the science of anthropology confirms the existence of *Homo sapiens* on Earth.[7] In the same way, the empirical science of Exopolitics and the law of evidence can provide *prima facie* evidence for the existence of intelligent life in the time-space dimension on Mars, including a human civilization on Mars.[8]

Exophenotypology of Martian Humanoids: Eyewitness Evidence

Former US chrononaut Andrew D. Basiago, whom I have interviewed many times since the year 2000 on his research into

---

[6] Slick. "Extraordinary claims require extraordinary evidence," http://bit.ly/1aAH796.
[7] Smithsonian Institution, "What does it mean to be human?" http://bit.ly/GYzJfW.
[8] Webre, "New data, law of evidence," http://bit.ly/P9nxw6.

intelligent life on Mars, is a Cambridge-trained environmentalist and attorney at law. He was one of America's early time-space explorers and the first American child to teleport. Mr. Basiago served in DARPA's Project Pegasus from 1968 to 1972, at ages 7 to 11, was involved in eight different types of time travel that had only recently been achieved by the US defense-technical community, and served in a second secret US space project, the CIA's Mars jump-room program, from 1980 to 1984, during his college years at UCLA. In that capacity, he made about 40 trips to Mars via jump-room as a US chrononaut.[9]

The first formal Exophenotypology of intelligent humanoid beings on Mars is based on eyewitness evidence from both Project Pegasus and the CIA jump-room program, as well as on photographic evidence from NASA's Mars rovers. Basiago developed the Exophenotypology of Martian humanoids from direct eyewitness observation of the *Homo martis terris* Exophenotype on Earth during his service as a participant on Project Pegasus, as well as from direct observation of the *Homo Martis martis* and *homo martis extraterrestrialis* types as a participant on the CIA Mars jump-room program on trips to Mars. Basiago also benefited from other accounts of Martian humanoid Exophenotypes, including official briefings by the CIA in preparation for Mars visits, and informal briefings with CIA officers, jump-room project personnel, and participants both on Earth and on Mars. His analysis of Martian humanoid Exophenotypes, as set out in his *The Discovery of Life on Mars*, was also part of his process of developing the Exophenotypology of Martian humanoids.[10]

Basiago writes:

There are three principal typologies of humanoid beings on Mars. [The first,] *Homo martis terris*, are our genetic relatives from the time

---

[9] Basiago, Project Pegasus. http://on.fb.me/19pjqmo.

[10] Basiago, Discovery. http://bit.ly/1837faW [This link will download the PDF files of his paper directly to your computer.]

before the solar system catastrophe of 9,500 BC, when Earth and Mars were in contact. They resemble bald, homely Earthlings. This is the type that my father and I met at Curtiss-Wright in 1970. They are human beings, like human beings from Earth. It was clear that by 1970 they both had a more advanced space-faring capability than we did on Earth and were working with the United States government.

[The second type,] *Homo martis martis*, are the indigenous Martian humanoids. They have narrower heads, pointier ears, longer fingers, and smaller bodies than Earthlings. They resemble the creature in the vintage film *Nosferatu*. We were advised that if hungry enough, they would sometimes kill and eat human visitors from Earth, so we should be wary of them. This is the type that I encountered when walking through the dilapidated brick city to deliver the data disk to the communications center there. They gave rise to the notion of "Martian" in folklore and popular culture on Earth.

[The third type,] *Homo martis extraterrestrialis*, resemble the Grey ETs of the UFO contact literature. This is the type that Courtney M. Hunt and William B. Stillings and I of the CIA's jump-room program encountered after I saw one that was sitting on the roof of the jump-room facility called The Corkscrew when we arrived on Mars one time in 1981-82. As I exited The Corkscrew ahead of my fellow chrononauts, I called out to them, "Court, Brett! A Grey. On the roof. Observing us!" They are probably the result of a branch of the Greys that was left back on Mars at some time in Martian history. It was this type of Martian that is depicted in my historic image, "The Humanoid Being on Tsiolkovski Ridge," which I published in 2008 in my landmark paper *The Discovery of Life on Mars*, which is the first image of a humanoid being on another planet ever published on Earth.[11]

There is a fourth typology of humanoid beings on Mars, the tiny ones that Dave Beamer has detected being crushed under the wheels of NASA's Mars Exploration rovers. I don't include this type of Martian humanoid in this taxonomy because they are far smaller than the three other typologies of Martians and us.[12]

---

[11] Basiago, Discovery. http://bit.ly/1aALaSK

[12] Basiago, "Three principal Exophenotypes of humanoid beings on Mars.' http://bit.ly/1cahkeq

Eyewitness Evidence of *Homo martis terris* Exophenotype

Basiago has publicly confirmed that in 1970, in the company of his late father, Raymond F. Basiago, an engineer for the Ralph M. Parsons Company who worked on classified aerospace projects, he met three astronauts of the Martian human civilization at the Curtiss-Wright Aeronautical Company facility in Wood Ridge, New Jersey, while the Martians were there on a liaison mission to Earth and meeting with US defense-technical personnel. These astronauts were of the *Homo martis terris* Exophenotype. According to Mr. Basiago, the astronauts asked to meet with him in the company of his father. Basiago states that the men resembled homely bald Earth humans, congruent with the Exophenotype.[13]

There are corroborative witnesses to Basiago's participation in time travel and teleportation in Project Pegasus, 1968-72. US Army Captain Ernest Garcia, whose career in US intelligence included serving both as a guard on the Dead Sea Scroll expeditions of Israeli archaeologist Yigal Yadin and as the Army security attaché to DARPA's Project Pegasus, confirmed Basiago's participation as a US chrononaut in Project Pegasus. Dr. Jean Maria Arrigo, an ethicist who works closely with US military and intelligence agencies, has also confirmed Basiago's participation in Project Pegasus.[14]

Martian Humanoid Exophenotypes: US CIA Mars Jump-Room Program

Basiago reports observing two of the three principal Exophenotypes of Martian humanoids while on Mars during trips in service on the CIA Mars jump-room program. These two are *Homo martis martis* and *Homo martis extraterrestrialis*.

---

[13] Webre, "New data, law of evidence," http://bit.ly/P9nxw6

[14] Ibid.

Q [Mike Spent]: What did you see while on Mars? What is Mars like? Is it inhabited?

A [Andrew Basiago]: Mars has a desert environment with scant water, sparse vegetation, and about 30 major land species. It has a blue sky and soils that range from light pink to rust red. It has three major humanoid typologies, namely, *Homo martis terries*, who are the descendants of the Earthlings who were left on Mars after the solar system catastrophe of 11,500 years ago [9500 BC]; *Homo martis martis*, who are the indigenous Martians that have bulbous heads, long faces, pointy ears, long fingers, spindly bodies, and fearful personalities; and *Homo martis extraterrestrialis*, who are a branch of the Grey ETs that are adapted to terrestrial life on Mars. During my 40 or so jumps, I encountered several animal species, including a predator that had a head like a Tyrannosaurus rex and a supple body, and a plesiosaur that was also a predator, with teeth running all the way down its throat. I met one of the *Homo martis martis* type of humanoid when he met Courtney M. Hunt and I at the jump-room facility called the Corkscrew and took us on a tour of his underground civilization, and I would see both adult and children of this type when I would walk through a dilapidated brick city to deliver a data disk to a telecommunications post that we had established there. Once, I saw one of the *Homo martis extraterrestrialis* species when I was exiting the Corkscrew with Courtney M. Hunt and William B. Stillings soon after arriving on Mars via jump-room. It was lurking atop the jump-room facility.[15]

The CIA's Mars jump-room program (1980-84) consisted of "jump-room" teleportation that had reportedly been designed and given to the US government by a specific species of the Grey Exophenotype hyperdimensionals, with jump-room facilities in New York City and Los Angeles, California, and a mission-control facility in Ohio. Some participants in the jump-room program still differ about its exact nature and destination.

---

[15] Spent, "Mars: Andrew D. Basiago," http://bit.ly/1aGBdTK. Two former participants in the jump-room program, former US chrononaut William B. Stillings and former Presidential adviser Bernard Mendez, have provided me with independent confirmation of Basiago's participation in the CIA Mars jump-room program.

Participants in the CIA Mars jump-room program or in US secret teleportation facilities to Mars have included some notable figures in US public life.

Q [Mike Spent]: Who else was involved in the trips? Can you name them all?

A [Andrew D. Basiago]: I can't name them all. I have, however, identified 17 Americans associated with the CIA's Mars jump-room program. Those who are known to have taken the jump-room to Mars are:

Andrew D. Basiago, former US Chrononaut
Raymond F. Basiago, Ralph M. Parsons Company
Major Ed Dames, US Army
Regina E. Dugan, future director of DARPA
Mary Jean Eisenhower, future People to People International, granddaughter, Dwight D. Eisenhower
Courtney M. Hunt, CIA officer
Linda Hunt
William C. McCool, future US Shuttle Columbia astronaut
Bernard Mendez, US Presidential special assistant
Barack H. Obama, future US President
William B. Stillings, former US chrononaut
Stansfield T. Turner, CIA Director
Michael C. Relfe, US armed forces security on Mars,
Arthur Neumann, physicist, Lawrence Livermore Laboratory [teleported to Mars in Lawrence Livermore Laboratory project]

Those who participated but did not take the jump-room to Mars are:
Stanley A. Dunham, mother of Barack Obama
Thomas J. Stillings, father of William B. Stillings
Michael Strickland[16]

---

[16] Spent, op. cit.

EVALUATING THE WITNESS EVIDENCE FOR INDIGENOUS MARTIAN INTELLIGENT LIFE

The eyewitness evidence of Andrew D. Basiago about indigenous Martian intelligent life, both on the surface and under the surface of Mars, is highly credible. I have had the opportunity to interview, question, and cross-examine Mr. Basiago closely about all aspects of his accounts regarding Project Pegasus and Mars since the year 2000. His verbal and written accounts have been consistent and verified by available witnesses and by documentary evidence, including written versions of his evidence that he has made available to me.

Basiago is extremely conscientious in the execution of his duties as an officer of the court and a member of the bar of the State of Washington. He consistently tells the truth—even punctiliously so—and is a conscientious reporter of his activities. The professional penalties for an officer of the court who engages in public deception or fraud regarding an issue of public importance, such as personal knowledge of indigenous Martian intelligent life, could be grave.[17]

Former US chrononauts William B. Stillings and Bernard Mendez, each a credible whistleblower eyewitness to intelligent life on Mars, have made their testimonies public at considerable personal risk. The testimonies of these three men about their roles as fellow participants in the 1980s CIA jump-room program are mutually consistent and congruent.[18]

EXOPHENOTYPOLOGY OF MARTIAN HUMANOIDS: DOCUMENTARY (PHOTOGRAPHIC) EVIDENCE

The Mars rover program has provided the historic opportunity to

---

[17] Webre, "New data, law of evidence," http://bit.ly/P9nxw6.

[18] Webre, "Mars visitors," http://bit.ly/10L2P5t; Webre, "Third whistleblower," http://bit.ly/O5TRcO.

photograph the surface of Mars. These photographs are transmitted to Earth and publicly released by NASA on the Internet. The public, including networks of Mars anomaly researchers, are then free to examine these NASA photographs for images of life forms, including signs of current or past intelligent civilization, such as cities, structures, statues, sculptures, etc.[19] The Mars Anomaly Research Society (MARS) is one such network of Mars anomaly researchers who have published images of Martian humanoids and other Martian life forms that have been located within NASA Mars rover photographs.[20]

Images of Martian life forms, including each of the three Exophenotypes of Martian humanoids, were first publicly disclosed by Basiago in his *The Discovery of Life on Mars*.[21] It accurately portrayed the Martian humanoid Exophenotypes, as later confirmed by life-long CIA employee Virginia Olds.

> Q [Mike Spent]: Have you been contacted by the US government lately about your disclosing these secrets?
>
> A [Andrew D. Basiago]: Yes. After I published my landmark paper *The Discovery of Life on Mars* in 2008, I was contacted by a senior case officer of the CIA, who confirmed that my paper was highly accurate about life forms on Mars.[22]

The Discovery of Life on Mars

In *The Discovery of Life on Mars*, Basiago describes the *Homo martis extraterrestrialis* Exophenotype as photographed by NASA rover Spirit:

> Similar humanoid beings are present at other locations in this photograph taken on Mars. Inside the Rock Enclosure on

---

[19] NASA, Mars Exploration Program, http://1.usa.gov/16wjz3Z.

[20] The Mars Anomaly Research Society (MARS), www.projectmars.net.

[21] Basiago, http://bit.ly/HbmVDJ.

[22] Spent, op cit.

Tsiolkovski Ridge, for example, a humanoid with two arms and two legs can be seen. He is kneeling with his back to Spirit. He has a bulbous head and an elongated body. He is wearing pants and a belt but is bare-chested. The scapulae in his upper back are evident. He is leaning over the wall away from the viewer. He may be reaching for, or lifting, a Martian child over the wall created by the back rock.

The humanoid being inside the Rock Enclosure on Tsiolkovski Ridge is easy to find in the photograph. Find the square in the upper left corner of the left solar array of Spirit, count five solar panels to the right on the edge of the array facing the valley, and proceed straight up the photograph. The C-shaped formation on the ridge is blue-gray in color. The humanoid is kneeling inside the enclosure of rocks. He is tiny and thin, like a praying mantis hatchling. He is interacting with other humanoids just beyond the rocks.[23]

The photographs Basiago is referring to are enlarged sections of NASA rover Spirit photographs PIA10214 and PIA11049. Basiago writes, "There is life on Mars. Evidence that the Red Planet harbors life and has for eons was discovered by the author by examining NASA photograph PIA10214, a westward view of the West Valley of the Columbia Basin in the Gusev Crater that was taken by the Mars Exploration Rover Spirit in November 2007 and beamed back to the Earth."[24]

Photographs of Martian Humanoid Exophenotypes

It may be useful to the reader to examine photographs of Martian humanoid Exophenotypes that have been discovered by Mars Anomaly Research Society (MARS) founder Andrew D. Basiago and MARS researcher Patricio Barrancos of Argentina.[25]

---

[23] Basiago, Discovery, p. 6. http://bit.ly/1aALaSK.

[24] Ibid., p. 1..

[25] For a selection of images, please see "Martian Humanoid Exophenotypes," Slideshow. http://bit.ly/19Noxgb.

Please understand that NASA intentionally makes the images of Martian humanoid Exophenotypes and other life forms difficult to perceive and recognize in NASA rover photographs of the surface of Mars such as NASA photo PIA10214.[26] According to Basiago, "[MARS] has catalogued seven techniques that NASA is using to distort its data," as follows.

1. Scale. Compress vertical and horizontal dimension to make everything look ordinary.
2. Contrast. Set light-and-dark ratio to extreme value or invert light and dark aspects altogether.
3. Color. Add false hues and unnatural colors.
4. Skew. Move forms out of vertical alignment.
5. Consistency. Paint over to make a solid color look black, white, or opaque.
6. Integrity. Alter natural forms to look like forms that do not occur naturally.
7. Content. Embed content with altered contrast, non-human data, and strange pixels.[27]

At my public presentations of Martian Humanoid Exophenotypes, about 75 percent of the audience can recognize and see the Exophenotypes of humanoids as well as the photographs of animals, statues, and structures on Mars that are shown (we poll the audience). At the November 2009 Barcelona Science and Spirit Conference, 50 percent of the audience that was polled could see. More recently, at public presentations in 2013, we have occasionally had up to 90 percent of the audience be able to perceive the Exophenotype of *Homo martis extraterrestrialis* in a

---

[26] NASA, Photojournal, PIA10214, Spirit's West Valley Panorama, http://1.usa.gov/18Ue2Iu.

[27] Basiago, "Hoagland affirms belief that no evidence exists Mars is inhabited," http://bit.ly/16yOv9f.

Mars Anomaly Research Society (MARS) photograph derived from a NASA Mars rover photograph of the surface of Mars.

REASONABLE IMPLICATIONS OF *PRIMA FACIE* EVIDENCE

Eyewitness evidence

There is ample *prima facie* eyewitness evidence for the existence of intelligent Martian humanoid life. For the sake of convenience, this evidence is set out in articles listed in "Sources and Resources" at the end of this book.[28] This evidence meets the standard of reasonability and is a *prima facie* demonstration of the existence of contemporary indigenous Martian humanoid life. There is *prima facie* eyewitness evidence to support a conclusion that a subsurface contemporary indigenous intelligent Martian human civilization exists [*Homo martis terris*], and that two other humanoid species exist [*Homo martis martis* and *Homo martis extraterrestrialis*].

Documentary (photographic) evidence

Likewise, there is ample *prima facie* documentary photographic evidence of intelligent Martian humanoid life, based on photographs from NASA rovers.[29] *Prima facie* evidence demonstrates that Mars has a fragile, post-collapse ecology, inhabited by at least three species of intelligent non-Earthling humanoid life. As concluded by Basiago, the *prima facie* evidence demonstrates that the Mars is inhabited by humanoid beings like and unlike Earthling humanity.

---

[28] Most of my articles in the "Sources and Resources" list present this evidence; I choose not to duplicate that list here and refer the reader to "Sources and Resources."

[29] These can be found in Basiago, Discovery. http://bit.ly/1aALaSK; Mars Anomaly Research Society (MARS), www.projectmars.net; and "Martian Humanoid Exophenotypes," Slideshow, http://bit.ly/19Noxgb.

US National Security Council and CIA Mars Jump-Room Program

The investigative process by which the secret CIA Mars jump-room program resulted in the development of a Martian Exophenotypology and became public knowledge is worth examining in its historic context, especially since the process of investigating the CIA Mars jump-room program occurred during the administration of a sitting US President, Barack H. Obama, who was himself, under his legal name Barry Soetoro, a CIA Mars jump-room participant in the early 1980s.[30]

The CIA jump-room started to become public knowledge when two former participants in the CIA's Mars visitation program of the early 1980s, Andrew D. Basiago and fellow chrononaut William B. Stillings, who was tapped by the Mars program for his technical genius, publicly confirmed that US President Barack H. Obama was enrolled in their Mars training class in 1980 and was among the young Americans from the program whom they later encountered on the Martian surface after reaching Mars via the jump-room, during visits to rudimentary US facilities on Mars that took place from 1981 to 1983. Their astonishing revelations provide a new dimension to the controversy surrounding President Obama's background and pose the possibility that it is an elaborate ruse to conceal Obama's participation as a young man in the US secret space program.

Mars training class held for future Mars visitors

According to Basiago and Stillings, in summer 1980 they attended a three-week factual seminar about Mars to prepare them for trips that were then later taken to Mars via teleportation. Remote-viewing pioneer Major Ed Dames, who was then serving as a scientific and technical intelligence officer for the US Army, taught

---

[30] Webre, "Mars visitors," http://bit.ly/10L2P5t.

the course. It was held at the College of the Siskiyous, a small college near Mt. Shasta in California. They state that ten teenagers were enrolled in the Mars training program. In addition to themselves, two of the eight other teenagers in Major Dames' class that they can identify today were Barack Obama, who was then using the name "Barry Soetoro," and Regina Dugan, whom Mr. Obama appointed as the 19th director and first female director of the Defense Advanced Research Projects Agency (DARPA) in 2009.

As many as seven parents of the ten students, all with ties to the CIA, audited the class. They included Raymond F. Basiago, an engineer for the Ralph M. Parsons Company who was the chief technical liaison between Parsons and the CIA on Tesla-based teleportation; Thomas Stillings, an operations analyst for the Lockheed Corporation who had served with the Office of Naval Intelligence; and Mr. Obama's mother, Stanley Ann Dunham, who carried out assignments for the CIA in Kenya and Indonesia. From 1981 to 1983, the young attendees then went on to teleport to Mars via a jump-room located in a building occupied by Hughes Aircraft at 999 N. Sepulveda Boulevard in El Segundo, California, adjacent to the Los Angeles International Airport.

Obama identified as having visited Mars at least twice

Basiago and Stillings have each issued public statements confirming that they both attended Mars training with Mr. Obama and later encountered him on Mars during separate visits.

On August 21, 2011, Basiago stated, "Something highly significant has happened, and that is that two individuals from the same Mars training class in 1980 (Basiago and Stillings) have met and are comparing experiences and are able to corroborate not only that they were on the surface of Mars together but that before reaching Mars via jump-room they were trained with a group of teenagers that included the current President of the United States (Obama) and director of DARPA (Dugan)."

Stillings' statement, released at the same time, read, "I can confirm that Andrew D. Basiago and Barack Obama (then using the name "Barry Soetoro") were in my Mars training course in Summer 1980 and that during the time period 1981 to 1983, I encountered Andy, Courtney M. Hunt of the CIA, and other Americans on the surface of Mars after reaching Mars via the jump-room in El Segundo, California."

In a statement made Sept 20, 2011, Basiago confirmed Mr. Obama's co-participation in the 1980 Mars training class, stating, "Barry Soetoro, a student at Occidental College, was in my Mars training class under Major Ed Dames at the College of the Siskiyous in Weed, California, in 1980. That fact has been corroborated by one of my other classmates, Brett Stillings. Two years later, when he was taller, thinner, more mature, a better listener, using the name 'Barack Obama,' and attending a different college, Columbia University, we crossed paths again in Los Angeles and I didn't recognize him as the person that I had been trained with in the Mars program and encountered on the surface of Mars. In fact, doing so would have been virtually impossible in any case, because measures had been taken to block our later memories of Mars shortly after we completed our training in 1980."

Basiago states that during one of his trips to Mars via jump-room that took place from 1981 to 1983, he was sitting on a wall beneath an arching roof that covered one of the jump- room facilities as he watched Mr. Obama walk back to the jump-room from across the Martian terrain. When Mr. Obama walked past him and Mr. Basiago acknowledged him, Mr. Obama stated, with some sense of fatalism: "Now we're here!"

Stillings states that during one of his visits to Mars, he walked out of the jump-room facility and encountered Mr. Obama standing beside the facility by himself staring vacantly into a ravine located adjacent to the facility.

Basiago thinks that it is virtually certain that Ms. Dugan also went to Mars, because he once encountered her at the building in El Segundo where the jump-room to Mars was located, as he was entering the building to jump to Mars and she was exiting it. "I know you!" she said, greeting him as she passed him in the lobby of the building.

STRANGERS IN A STRANGE LAND

Basiago, Obama, Stillings, and Dugan went to Mars at a time when the US presence on Mars was only just beginning but many had already gone. Basiago states that in the early 1980's, when they went, the US facilities on Mars were rudimentary and resembled the construction phase of a rural mining project. While there was some infrastructure supporting the jump-rooms on Mars, there were no base-like buildings like the US base on Mars first revealed publicly by Command Sgt. Major Robert Dean at the European Exopolitics Summit in Barcelona, Spain, in 2009. The primitive conditions that they encountered on Mars might explain the high level of danger involved. Basiago and Stillings agree that Major Dames stated during their training class at the College of the Siskiyous in 1980, "Of the 97,000 individuals that we have thus far sent to Mars, only 7,000 have survived there after five years."

In light of these risks, prior to going to Mars, Basiago received additional training from Mr. Hunt, a career CIA officer, who showed Basiago how to operate the respiration device that he would wear only during his first jump to Mars in July 1981, provided him with a weapon to protect himself on Mars, and took him to the Lockheed facility in Burbank, California, for training in avoiding predators on the Martian surface.

When they then first teleported to Mars in Summer 1981, the young Mars visitors confronted the situation that Major Dames had covered at length during the class the previous summer, that one of their principal concerns on Mars would be to avoid being devoured by one of the predator species on the Martian surface, some of which they would be able to evade, and some of which were impossible to evade if encountered.

The Mars program was launched, Basiago and Stillings were told, to establish a defense regime protecting the Earth from threats from space and, by sending civilians, to establish a legal basis for the US to assert a claim of territorial sovereignty over Mars. In furtherance of these goals and the expectation that human beings from Earth would begin visiting Mars in greater numbers, their mission was to acclimate Martian humanoids and animals to their presence or, as Major Dames stated during their training, "Simply put, your task is to be seen and not eaten."

It is not known whether NASA's Jet Propulsion Laboratory, which is located in Pasadena, California, had a hand in selecting the young people for their dangerous interplanetary mission to Mars, but it is conspicuous that all four had Pasadena connections. Basiago's father worked for the Ralph M. Parsons Company, headquartered in Pasadena. Stillings was residing in La Canada, a suburb of Pasadena. Mr. Obama had just completed a year of undergraduate studies at Occidental College in Eagle Rock, near Pasadena. Ms. Dugan was attending the California Institute of Technology, located in Pasadena.

1980: US Secret Mars Teleportation Program and Rudimentary Facilities on Mars

The firsthand, eyewitness testimony of Basiago and Stillings about the existence of a secret US presence on Mars made possible by a revolutionary jump-room technology that has been concealed from the public is congruent with similar accounts given by three other Mars whistleblowers:

- Former US serviceman Michael Relfe, who spent 20 years as a member of the permanent security staff of a US facility on Mars;

- Former Lawrence Livermore National Laboratory scientist Arthur Neumann, who has testified publicly that he teleported to a US facility on Mars for DoD project meetings; and
- Laura Magdalene Eisenhower, great-granddaughter of US President Dwight D. Eisenhower, who in 2007 refused a covert attempt to recruit her into what was described to her as a secret US colony on Mars.

Mars researchers, including physicist David Wilcock, estimate that as a result of the jump-room technology that Relfe, Basiago, Neumann, and Stillings have described, the US colony on Mars that Eisenhower was invited to join might number 500,000 individuals. With multiple whistleblowers coming forward and corroborating each other's testimony, it now seems inevitable that both the cover-up of the US presence on Mars and Mr. Obama's personal involvement in it will soon become matters of great public interest.

On January 3, 2012, Tommy Vietor, the official spokesperson of the US national security council, a body that includes the US President, Vice President, Secretary of Defense, Secretary of State, and National Security Adviser, denied that Obama had participated in the CIA Mars jump- room program. According to Huffington Post UK reporter Michael Rundle,

> The White House has denied that President Barack Obama was teleported to Mars in the 1980s as a member of a secret CIA project hosted at a community college in California. Two people have made the claim while asserting that they themselves were time-traveling government agents sent across the solar system to explore new worlds. Andrew D. Basiago and William Stillings have argued that they served as "chrononauts" at the request of the Defense Advanced Research Projects Agency. The two men also insist that the Commander in Chief was part of the team and made two trips to the Red Planet via a "jump-room." At that time, aged 19, Barack was reportedly known as Barry Soetoro....
>
> Stillings made a statement in August 2011 that said: "I can confirm that

> Andrew D. Basiago and Barack Obama (then using the name "Barry Soetoro") were in my Mars training course in Summer 1980 and that during the time period 1981 to 1983, I encountered Andy, Courtney M. Hunt of the CIA, and other Americans on the surface of Mars after reaching Mars via the jump-room in El Segundo, California."

Now *Wired* magazine reports that the White House has been forced—albeit with tongue firmly in cheek—to deny that Obama ever made the 200 mile trip. So has Obama been to Mars? "Only if you count watching Marvin the Martian," Tommy Vietor, the spokesman for the National Security Council, told *Wired*.[31]

A Third Whistleblower

Following the January 2012 White House denial, the next phase of the CIA's Mars jump-room becoming public knowledge began with the appearance of Bernard Mendez to corroborate the existence of the program and Basiago's accounts of it.[32]
According to Bernard Mendez' biography,

> Bernard Mendez was born into a family involved deeply in military service. His father, Bernard Mendez, Sr., served in World War II for the US Army and was intimately involved in secret military projects.
>
> At the age of two, Mendez was observed by family members to have a seeming ability to see objects in the invisible range of the spectrum. Upon medical examination, it was discovered that he possessed an unusually high number of rods and cones in his eyes that allowed him to see in the dark.
>
> Throughout his youth, Mendez had numerous direct encounters with several ET races, most of which were Greys. This continued into his teenage years.
>
> By the age of 16, Mendez had enlisted into the United States Navy, where he was assigned to the Special Forces command. On October 9, 1970, he was part of a very bizarre abduction incident in Australia

---

[31] "White House denies President Obama travelled to Mars via teleport at age 19," http://huff.to/1fU5xla.
[32] Webre, "Third whistleblower," http://bit.ly/O5TRcO.

and [was] eventually rescued by Navy Seals, as reported in the Associated Press. This incident caught the attention of then-President Richard Nixon, who asked to meet the young sailor at the White House.

Eventually, Nixon recruited Mendez to be a special assistant in a secret government program dealing directly with extraterrestrials. He was sent to locations all over the world, including Alaska, Australia, China, France, Antarctica, the North Pole, Catalina Island, K2, the Fjords, as well as the infamous secret underground bases known as Area 51 and Camp Hero (aka Montauk Naval Air Station), Groom Lake, China Lake and Clove Lakes. He visited these places in order to negotiate directly with various ET representatives on behalf of the US government. Mendez even visited a secret underground base in England known as the "Bat Cave" and the former Soviet Union's version of NASA, Kapustin Yar, near Volgograd, Russia.

During his years of service, Mendez obtained what is known as a cosmic-level clearance, designated "Yankee White," which permitted him to work for the White House. He rose through many different access levels ... before eventually reaching Access Level 59. He was told he would go no further even though there was an Access Level 60, which was the highest level.

From Spring 1971 until the Watergate scandal forced President Nixon to resign on August 9, 1974, Mendez worked for the Nixon White House as a go-between with various Greys and Pleiadians. He continued in secret government service for a brief period of time into the 1980's before retiring at a young age.

In 2011, Mendez heard an interview of Andrew D. Basiago on talk radio's Coast to Coast AM with George Noory about his participation in a secret government time travel program called Project Pegasus in the early 1970's and a later secret Mars visitation program in the early 1980's. Mendez contacted the American chrononaut and they confirmed that they had served together in Project Mars, which was the CIA's Mars jump-room program of the early 1980's, with Barack H. Obama, Regina E. Dugan, William B. Stillings, William C. McCool, and other young Americans of their generation.[33]

---

[33] Mendez, http://on.fb.me/1adQyyt.

Confirmation of CIA Jump-Room Program

Bernard Mendez became the third whistleblower to confirm that US President Barack H. Obama was a participant in the CIA's teleportation jump-room program of the early 1980s. He made this allegation during a seminar held in Vancouver, British Columbia, on June 1, 2012, under the auspices of the ExoUniversity.org. During the seminar, Mendez stated that he, Mr. Obama, former DARPA director Regina E. Dugan, Andrew D. Basiago, and William B. Stillings attended a jump-room training class taught by Maj. Ed Dames at College of the Siskiyous in summer 1980. He also stated that he teleported with Mr. Obama and Mr. Basiago in jumps that took place from 1981 to 1983, when Obama and Basiago were college students. At the time, Mendez, who once served as a special assistant to President Richard M. Nixon, was investigating the jump-room program for the US intelligence community.

On November 8, 2011, I was the first to report revelations made by Basiago and Stillings that Obama had served with them in the jump-room program of the early 1980s. The two chrononauts then appeared on late-night talk radio's Coast-to-Coast AM with Laura Magdalene Eisenhower, the great-granddaughter of US President Dwight D. Eisenhower, to discuss their involvement in the US secret space program. Major Dames called the show to deny his involvement in the jump-room program.

Then, in January 2012, through National Security Council spokesperson Tommy Vietor, Mr. Obama also denied that he was part of the 1980-83 CIA jump-room program. Mendez' statements confirming the existence of the CIA's jump-room program and the roles played by Major Dames and Mr. Obama in it directly refute these denials. Other jump-room program participants identified by Mendez and Basiago have neither confirmed nor denied their involvement in the program as of this writing.

Asked for comment on Obama's silence in the face of confirmation of his participation by three jump-room participants,

Basiago stated, "President Obama campaigned for president pledging that his administration would be one of transparency. He must now fulfill that promise and acknowledge that he served with us in the jump-room program."

During the ExoUniversity.org seminar, Mendez explained that his mission to evaluate the CIA's jump-room program originated from discrepancies being reported by the jump rooms on the US east and west coasts. The east coast jump room was located in New York City, the west coast jump room in El Segundo, CA.

According to Mendez, the jump-room technology had been transferred from a species of Grey extraterrestrials to the US government. The west coast and east coast jump rooms were reporting up to 40 incidents of participant injuries per month occurring during jump-room "teleportation" to unknown environments in space. Yet, several days after each teleportation jump, the reported injuries to participants had disappeared.

Both the east coast and west coast jump rooms experienced losing power at times when their control room in Ohio was reporting normal operation of both jump rooms. The speculation was that the source of the jump-room technology (Grey extraterrestrials) may have been covertly interfering with the functioning of the jump-room teleportation technology, unbeknownst to control-room operators in Ohio. The US government deployed Mendez, along with an evaluation team that included several prominent US astronauts, to find out the causes of these discrepancies and the true destinations of the jump rooms. Basiago and Stillings have confirmed that Mr. Mendez was a federal investigator whose primary function was investigating questions like where the jump rooms were going.

SYNTHETIC QUANTUM ENVIRONMENTS (SQES)

Upon arriving at the jump-room training class at College of the Siskiyous in 1980, Mendez debriefed Major Dames about his intelligence mission to evaluate the program. He then began a series of test jumps from the west coast jump room. On one such test jump he teleported with Mr. Obama and Mr. Basiago to a planetary environment that he initially thought might be the Mars they had been trained for. He recalls that when he had Mr. Obama shoot a flare at the sky, the flare bounced off a ceiling at about 62 feet up, indicating that they had teleported to an artificial environment or domed enclosure of some kind. The precise location of that enclosure was uncertain, and may have been on a celestial body, such as Mars, or a freestanding holographic creation in space.

Mendez determined after multiple jumps that the destination the jump rooms were reaching was not Mars, which was out of position when most of the jumps took place. Its atmosphere lacked enough oxygen to sustain human life. Gravity on the surface was stronger than Mars' gravity would have been. The temperature swings on Mars (from -175°F to +225°F) would have made surface visits not survivable. Mars' lack of an ionosphere would have led to the irradiation of the jumpers. He concluded that the jump rooms were actually taking the participants to a "synthetic quantum environment" or SQE, a term invented by Basiago to describe a fold in the time-space continuum in which the Grey extraterrestrials have established an "artificial holographic planetary domain."

During the ExoUniversity.org seminar, Basiago stated: "According to Bernard Mendez, the jump rooms were not visiting Mars but 'folds' in time-space created by the Grey extraterrestrials that the Apollo astronauts called 'slots.' These 'slots' or niches in time-space contain synthetic quantum environments in which the Greys have stored event scenarios that have taken place at different times on different planets, including Earth and Mars."

The jump-room program resulted, Mendez explained, from an ET-human liaison project involving the Greys and the US government, in which the young Americans involved were selected by the Greys and then trained by the CIA. "This astonishing revelation means that President Barack Obama is a contactee selected by the Grey extraterrestrials to participate in the secret space program," Basiago commented.

Dimensional Ecology: New Land in Time-Space?

According to Mr. Mendez, the US government has identified 153 "synthetic quantum environments" constructed by the Greys in the near-Earth environment, ranging 400 miles from Earth, on the surface of the Earth, and beneath the surface of the Earth. Because they contain new land that might be territorially acquired, the US government was actively discovering and exploring these SQEs with a view to making them part of the United States.

Mr. Mendez was tasked to negotiate with the Greys as part of his evaluation mission. During the negotiations, the Greys indicated that they had built the SQEs to be used for conditioning and educating humanity to be able to sustain a future role among organized intelligent civilizations in space and in the interdimensional multiverse.

A Chrononaut's Contrary View

In the panel discussion sponsored by ExoUniversity.org held on June 2, 2012 that included all three whistle blowers—Basiago, Mendez, and Stillings—Stillings shared his view that the destination the jump rooms were reaching was, in fact, Mars. He argued that the Martian moons Phobos and Deimos could be seen in their proper locations and trajectories in the sky and that jumper Courtney M. Hunt once stated that the Viking 2 apparatus was about 20 miles away from the jump-room infrastructure that the jumpers called "the Corkscrew" because of its spiral structure. These clues indicated that the jumpers were on the real Mars and not a simulation.

NASA cannot be relied on to provide accurate data about Mars, Stillings stated, citing the space agency's claim that atmospheric pressure on Mars is only eight millibars. Such an atmospheric pressure could not have sustained the parachute deployment of the Mars Exploration Rovers Spirit and Opportunity to the surface.

Basiago, who has spearheaded efforts to have America's chrononauts come forward and share their experiences with the public, called Mendez' account "a cosmic-level disclosure event." For his part, Basiago thinks that Mendez and Stillings may both hold a key to the truth. He thinks that since the origins and operation of the jump rooms were unknown, they might have sometimes been taking the jumpers to Mars, but at other times were being diverted by the Greys to SQEs where human responses to the threshold of contact with the wider universe were studied.

In support of this explanation, Mr. Basiago cited the fact that Grey observation of US activities in time-space was sometimes noted during time probes using US time-travel technologies while he was serving on DARPA's Project Pegasus as a child in the early 1970's. He also cited the fact that power losses were being recorded during times when the jump rooms were ostensibly functioning normally. This could be evidence that during operation, the Greys were diverting the jump rooms from Mars to elsewhere.

Basiago stated that since NASA has been lying about natural conditions on Mars, determining whether Mendez' "elsewhere" interpretation of the CIA jump-room program or Stillings' "Mars" interpretation is the correct one will hinge on official declassification. He observed:

> What we have now is three fellow jump-room participants sharing their experiences... We were trained for Mars in summer 1980 and the domain that we were visiting from 1981 to 1983 was certainly understood to be Mars. What we are grappling with now is Bernard's claim that as the US government investigator tasked to study the

project he discovered that we were not visiting Mars but a... simulation of Mars architected by the Greys in a bubble Universe... That's a highly provocative claim, but since Bernard Mendez and Brett Stillings and I were project participants together, in the interests of truth, I have facilitated all three of us coming forward publicly and sharing our understanding of what we were part of... Even if it was not Mars that we were visiting, all three of us agree that it was an off-planet location in time-space.

OBAMA AS A TIME-TRAVEL PRE-IDENTIFIED CIA ASSET

Whistleblower Bernard Mendez' detailed testimony that he not only participated in the 1980-83 CIA jump-room program with Barack Obama, but also engaged Mr. Obama in jump-room experiments, such as a flare experiment to determine the location that the jump-room technology was teleporting to, further bolsters emerging evidence that Barack H. Obama was involved at the early age of 19 in a highly classified, top-priority US national security project and went on to be a life-long CIA asset or operative. It is logical, then, that the Obama administration would use such a high-level asset as Tommy Vietor, official spokesperson of the National Security Council, to deny, on January 3, 2012, that Mr. Obama was a participant in the 1980s CIA jump-room program.

The CIA jump-room program, and its origin with a species of Grey extraterrestrials with whom the US government has had an ongoing secret human-extraterrestrial liaison program, is of the highest national security to the US government. The CIA and US national security apparatus surrounding President Barack Obama will do everything in its power to suppress evidence that Mr. Obama

- was part of a secret teleportation jump-room program using Grey extraterrestrial technology to reach SQEs, destinations architected by Grey extraterrestrials;

- had been since 1980 a full-time covert CIA operative, deployed to infiltrate, perform surveillance on, and report back on community activist, Leftist, African nationalist, and Islamic circles; and
- was pre-identified in 1971 via quantum-access time travel as a future US President and had been essentially briefed and groomed for the job by CIA since 1980.[34]

---

[34] Obama's identity as a CIA operative/asset and as being pre-identified as a future US President is explored in my "How intelligence legend & Manchurian candidate Barack Hussein Obama was created," which can be viewed online at http://slidesha.re/18Im2fE.

CHAPTER 4
HYPERDIMENSIONAL CIVILIZATIONS

Prima facie replicable empirical evidence supports the hypothesis of a dimensional ecology among the universes of the Exopolitical dimensions and the spiritual dimensions in the Omniverse. Hyperdimensional civilizations are active in the dimensional ecology in dimensions adjacent to our Earth's time-space dimension. This witness and documentary evidence indicates that the dimensional ecology is used by various species of intelligent hyperdimensional civilizations, each carrying out specific Exopolitical agendas with respect to humanity. For example, from the platforms of the dimensions in which they are based, various hyperdimensional civilizations reportedly carry out "interdimensional teleportation" on large numbers and categories of humans.[1]

There is *prima facie* evidence that our Earthling time-space dimension solar-system civilization is in a dimensional ecology with hyperdimensional intelligent civilizations. As with the time-space dimension solar-system intelligent civilizations on Mars,

---

[1] Webre, "Up to 1 billion humans are abducted," http://exm.nr/HfEpOC.

the empirical method and the law of evidence can demonstrate the existence of hyperdimensional intelligent civilizations in our universe in the Multiverse.

Hyperdimensional civilizations are intelligent civilizations based in dimensions in our universe parallel to our own time-space Earth dimension or in universes parallel to our universe. They can be based in higher dimensions of our own solar system, of our galaxy (the Milky Way) or of other galaxies of our universe, or in universes parallel to our own universe in the multiverse. They may use technologically advanced interdimensional transport when teleporting into our known physical universe or our Earth time-space dimension. They may also use advanced consciousness technologies that permit them to teleport interdimensionally.[2]

Eyewitness Evidence of Hyperdimensional Civilizations

As was mentioned in Chapter 2, there is replicable *prima facie* empirical evidence to show that a low estimate of 2 percent of the US population and a high estimate of 14.7 percent of the total human population may have been physically, etherically, or mentally temporarily or permanently teleported to spaceships or dimensions outside our time-space Earth dimension by hyperdimensional civilizations.[3]

The eyewitnesses to these teleportations report that they are carried out by various species of intelligent Grey hyperdimensional civilizations, as well as by other hyperdimensional civilizations. Expert estimates of the number of known Grey intelligent species range as high as 150 different species.[4] One study remarks on

---

[2] Webre, "Intention experiments, interdimensional UFOs, and multi-dimensional beings converge at Mt. Adams, WA," http://exm.nr/18DT1gZ.

[3] Webre, "Report: Between 2 percent," http://bit.ly/1f4GcCi.

[4] Webre, "Mary Rodwell," http://bit.ly/17tBY7B.

the "Hive Minds" of the [hyperdimensional] Greys and alleged abilities at "topological thinking" at speeds faster than the fastest computers known to mankind. The neurological superiority of Greys with discussion on the various cases, that specifically depict humans fearing Greys [per Dr. David Jacobs] and Greys that have worked lovingly and positively with contactees, such as Suzanne Hansen [per Mary Rodwell]; Greys who have been seen with Anunnaki such as the case involving Futant[5] and Contactee Neil Freer.[6]

Replicable empirical research published by the International Community for Alien Research (ICAR)[7] confirms that a hyperdimensional civilization consisting of amphibian reptilian hybrids and Grey cybernetic clones has reportedly abducted up to 14.7 percent of the present world population, or approximately 1 billion persons out of the current estimated human population of about 6,783,000,000 as of May 31, 2009. These human abductees, according to this replicable empirical research, have had their consciousness and/or their physical bodies taken by a quantum, hyperdimensional process and have then been returned to their original location (and in some cases original time) with erased memories of the abduction.[8]

One 2002 study concluded that 2 percent of the US population has been abducted by hyperdimensional civilizations. "In 1991, [Budd] Hopkins, [Prof. David] Jacobs, and sociologist Dr. Ron Westrum commissioned a Roper Poll in order to determine how many Americans might have experienced the abduction phenomenon. Of nearly 6,000 Americans, 119 answered in a way that Hopkins *et al.* interpreted as supporting their ET interpretation of

---

[5] According to Neil Freer, "Futant is a term coined by Timothy Leary, Ph.D., former Harvard lecturer in psychology and the engine behind the peaceful Jeffersonian revolution of the '60's. Melding future and mutant, it defines those 1.5-2 percent of the population at any given time who are genetically 'programmed' to be the evolutionary future scouts."

[6] Galactic Diplomacy: The Greys, http://bit.ly/181e3qD; Neil Freer, Sapiens Rising, http://bit.ly/19H6gTj

[7] International Community for Alien Research, ICAR, http://bit.ly/1aEN6v7

[8] Webre, "Report: Between 2 percent," op. cit.

the abduction phenomenon. Based on this figure, Hopkins estimated that nearly four million Americans might have been abducted by extraterrestrials."[9]

Memories of the hyperdimensional abduction retrieved through hypnotic regression and similar techniques reveal consistent patterns during the abductions, related to a covert agenda toward the human species by this hyperdimensional species. Because the abduction research data conflicts with the conventional paradigm of human reality, it is filtered out and denied, even by major researchers in the field of extraterrestrial studies. Cognitive dissonance or "the state of having inconsistent thoughts, beliefs or attitudes, especially as to behavioral decisions and attitude change" is one way of describing categories of responses to knowledge about and research into these hyper-dimensional abductions.

ALIEN ABDUCTIONS AS PSYCHOLOGICAL EVENTS?

The canon of conventional academic science is that alien abductions are a psychodynamic artifact. One source states: "The abduction phenomenon has garnered substantial attention from mainstream scientists and mental health professionals, who overwhelmingly doubt that the phenomenon occurs literally as reported and who have proposed a variety of alternate explanations, including '[d]eception, suggestibility (fantasy-proneness, hypnotizability, false-memory syndrome), personality, sleep phenomena, psychopathology, psychodynamics, [and] environmental factors.'"[10]

---

[9] Alien abduction claimants, The Roper Poll. http://bit.ly/195jNAh.
[10] "Alien abduction," http://bit.ly/HqHEmX.

MILABs

Military abductions (MILABs) are a form of psychological warfare and psyops abductions done by human military-intelligence agencies against targeted individuals in order to create false terror about hyperdimensional extraterrestrial civilizations, which are uniformly beneficent to humanity.

The ICAR research confirms that alien abductions are largely done by hyperdimensional civilizations, not by MILAB military abductions, which represent a secondary portion of alien abductions. ICAR's research conclusions are backed by extensive, replicable research methodology and data, including objective correlation of answers to direct questions asked of Grey and amphibian-reptilian hyperdimensionals about the nature and purpose of alien abductions and their agenda towards the planet and humanity, by groups of 20-50 recurring abductees.[11]

CASE STUDY: HYPERDIMENSIONAL WARFARE IN THE DIMENSIONAL ECOLOGY

A case study that illustrates how intelligent hyperdimensional civilizations strategize and operate within the dimensional ecology is that of hyperdimensional warfare. This case study, which also may define the critical limits on any official extraterrestrial "disclosure" or open contact with extraterrestrial or hyperdimensional civilizations, may be deduced from eyewitness whistleblower evidence that there may be a hyperdimensional armed conflict going on over control of our solar system.[12] If this apparent state of hyperdimensional "war" is factual (and not a False Flag operation) and is part of a universal and galactic energetic transformations, then "official" extraterrestrial

---

[11] Webre, "Report: Between 2 percent," op. cit.

[12] Webre, "When will extraterrestrial 'disclosure' or contact in the public domain happen?" http://bit.ly/1gdjmLC.

disclosure or peaceful open contact is likely to occur in the near future.

The eyewitness evidence for such a hyperdimensional "war" includes whistleblower testimony by two former US armed forces personnel at a secret US base on Mars. Their testimony describes actual scenarios in the dualistic military terms of defender and "enemy," or both sides of the dialectic of enemies known as the armed combat of war. As set out in the whistleblower eyewitness evidence (see below), the quantum-access nature of the "enemy's" arsenal of weapons systems suggests that the "enemy" may be a hyperdimensional or advanced third- or fourth-dimensional extraterrestrial civilization from another solar system or galaxy. The objective of this hyperdimensional "war" appears to be control of the third time-space dimension of our solar system.

WHAT IS A HYPERDIMENSIONAL "WAR"?

A conventional definition of war is

> a behavior pattern exhibited by many primate species including humans, and also found in many ant species. The primary feature of this behavior pattern is a certain state of organized violent conflict that is engaged in between two or more separate social entities. Such a conflict is always an attempt at altering either the psychological hierarchy or the material hierarchy of domination or equality between two or more groups. In all cases, at least one participant (group) in the conflict perceives the need to either psychologically or materially dominate the other participant. Amongst humans, the perceived need for domination often arises from the belief that an essential ideology or resource is somehow either so incompatible or so scarce as to threaten the fundamental existence of the one group experiencing the need to dominate the other group. Leaders will sometimes enter into a war under the pretext that their actions are primarily defensive; however, when viewed objectively, their actions may more closely resemble a form of unprovoked, unwarranted, or disproportionate aggression.
>
> In all wars, the group(s) experiencing the need to dominate other group(s) are unable and unwilling to accept or permit the possibility

of a relationship of fundamental equality to exist between the groups who have opted for group violence (war). The aspect of domination that is a precipitating factor in all wars, i.e., one group wishing to dominate another, is also often a precipitating factor in individual one-on-one violence outside of the context of war, i.e., one individual wishing to dominate another.[13]

Hyperdimensional "War" and Quantum-Access Technology

The possible hyperdimensional "war" has come to public light because of the operational use of quantum-access time-travel jump gates and quantum-access technology at the US secret base on Mars. Quantum-access technology is used to time-travel US armed service personnel who serve as permanent staff at that base and to age-regress them back to their original time-space starting date after completion of their 20-year rotation cycle on Mars. The base thus may be hyperdimensionally "cloaked" from civilian society on Earth, and hidden in covert niches in the time-space hologram. As described by former US chrononaut Andrew D. Basiago, one early application of time-travel technologies was the hiding of national security secrets from the Soviet Union during the Cold War, for example, by time-traveling these secret documents to quantum- access niches in the time-space hologram.[14]

Eyewitness Whistleblower Evidence of Warfare with Hyperdimensionals

Two independent whistleblowers, both such former US armed-service personnel stationed at the secret facility on Mars, have now come forth with detailed accounts of their experiences in warfare with hyperdimensional intelligent civilizations.

---

[13] War. http://bit.ly/16mWQKW.

[14] Basiago, Project Pegasus. http://bit.ly/1aUmkxp.

Michael Relfe is one such whistleblower. A former member of the US armed forces, he was recruited as a permanent member of the secret US facility on Mars. In 1976 (Earth time), he was teleported to the US Mars facility and spent 20 years as a permanent member of its security staff. In 1996 (Mars time), Mr. Relfe was time-traveled via teleportation and age-regressed 20 years, landing back at a US military base in 1976 (Earth time). Mr. Relfe then served six years in the US military on Earth before being honorably discharged in 1982.[15] He is a credible whistleblower whose life and career have suffered greatly because of his decision to inform the public of the secret war against hyperdimensional intelligent civilizations.[16]

In a two-volume book authored by his wife, Stephanie Relfe, *The Mars Records*, Relfe describes the two types of individuals at the secret Mars colony:

To clarify: Remember there are two kinds of people that I remember.

1. People visiting Mars temporarily (politicians, etc.) – They travel to and from Mars by jump gate. They visit for a few weeks and return. They are not time traveled back. They are VIP's. They are OFF LIMITS!!

2. Permanent staff – They spend 20 years' duty cycle. At the end of their duty cycle, they are age reversed and time shot back to their space-time origin point. They are sent back with memories blocked ... to complete their destiny on Earth.[17]

Relfe also discusses the presence and functions of Reptilian and Grey extraterrestrials at the secret Mars colony.

---

[15] Webre, "Basiago and Eisenhower reveal 'Marsgate' and make case for 'Alternative 4'," http://exm.nr/17ya8VD.

[16] Webre, "Time Travel and Political Control," http://bit.ly/LoBF0T.

[17] Relfe, The Mars Records, Vol. 2, p. 204.

EL: What about the Reptilians?

Michael Relfe: Yes. They are racially related (Draconians, Reptilians, Greys).

EL: Do any Greys and Reptilians live on the Mars Base?

Michael Relfe: Yes, some are stationed there. I remember the Greys as doctors or technicians. I believe the Reptilians stay camouflaged (cloaked) most of the time. They prefer to appear human because they are naturally fierce-looking.[18]

THE HYPERDIMENSIONAL "ENEMY" AT THE SECRET US MARS FACILITY

Relfe describes various scenes of armed conflict between the US military presence at the secret Mars base and an armed hyperdimensional "enemy." He also describes the weapons systems that the hyperdimensional "enemy" in this ongoing conflict possesses.

The following excerpt from the book *The Mars Records* illustrates the nature of the armed conflict occurring around the US secret base on Mars and the types of weapons used. It is taken from a discussion of "time-travel jump gates," that is, portals used for time-travel age regression of permanent staff of the base to return them to their original time-space point of departure from a US military intelligence base on Earth. This use of time-travel jump gates makes it possible to hide the US secret base on Mars in the time-space hologram from discovery by Earthling civilian society.

Relfe describes the secret Mars base as being under attack by a hyperdimensional "enemy" that apparently can time-travel and may be a third- or fourth-dimensional extraterrestrial civilization from another solar system or galaxy.

EL: Tell me about the time travel jump gates.

Michael Relfe (MR): I was not a jump gate technician so I only remember things from a layman's point of view. This technology was

---

[18] Ibid., p.205.

a result of the Philadelphia Experiment projects as described by Dr. Al Bielek. It's one of those things that are "Ohh, Wow, Fantastic" the first time you see them, and then they become taken for granted. I also remember that they were guarded REALLY tight and that every moment of use was accounted for and logged.

I remember that there were several jump gate stations on Mars Base. These stations were "hooked to" other places and they were defended against someone or something. I remember that RV technicians hooked to the machines assisted in that defense. Remember that Tactical Remote Viewing is not only used to terminate targets. Termination is a tiny percentage of all operations. It is used mostly to defend against the enemy. It is used to defend Very Important People (VIP's), Very Important Equipment (VIE's), areas and regions of the planet. In addition, some weapons systems require an RV operator to monitor and direct them. There are physical weapons systems (example like a particle beam projector) and there are other types of weapons that are energetic (Psi) types of systems. Also RV operators are commonly assigned to monitor attack craft on patrol.

The enemy also has Psi types of weapons systems that they direct against our physical type static shielding. Static shielding doesn't change. It stays the same no matter what you throw at it. To augment the static shielding, you have RV operators that monitor and probe the shields looking for penetration. If penetration is discovered, an RV operator can trace the source through space-time and deal with the intruder. RV operators on duty are not alone. They have backup operators in addition to supervisor operators, they can instantaneously get backup if the need arises. I have seen many ads for books and courses for remote viewing. Most of what passes for RV at this time is about the first 3 or 4 days of training at Mars Base. And most people involved in this sort of thing would, unfortunately never survive the training. The drugs would kill them quick.[19]

In one passage, Relfe recounts an "attack run" he undertook on "an enemy ship" where he lost consciousness when the enemy missile penetrated his ship. He "was flying solo in a space ship when he encountered an enemy craft. Instead of fleeing he went

---

[19] Ibid., vol. 2, p. 203.

for the enemy craft. It fired a missile at him, which did not explode his ship, but it did penetrate his ship and go through him. This caused incredible damage. He would have died had he not had a special suit which was programmed to administer first aid."[20] Relfe describes the advanced technologies possessed by the hyperdimensional "enemy."

> The [hyperdimensional] enemy has at their disposal multiple technologies that will allow them to abduct an individual from their bedroom, examine/operate/ indoctrinate them, and return them to their bedroom without being detected. They can "keep" an individual for weeks and return them to a time a few minutes after they were taken. The names of those technologies that I am aware of are:
>
> Jump gate technology
> Fractal Jump gate technology
> Transporter technology
> Fractal Transporter technology
> Teleportation technology
> Fractal Teleportation technology
> Time Travel technology
> Fractal Time Travel technology
> Dimensional Travel technology
> Fractal Dimensional Travel technology
> Wormhole Travel technology
> Fractal Wormhole Travel technology
> Resonance technology
> Fractal Resonance technology
> Magical technology
> Fractal magical technology
> Walk-through-the-walls technology
> Fractal walk-through-the-walls technology
> A "sender" sending something into your room
> A "sender" pulling you out of your room[21]

---

[20] Ibid., vol. 1, pp. 65, 77.

[21] Ibid., vol. 2, p. 110.

What is This Hyperdimensional "War" About?

The dimensional nature of the "enemy's" arsenal of weapons systems suggests that the "enemy" may be a third- or fourth-dimensional extraterrestrial civilization from another solar system or galaxy. There are other possible interpretations of the eyewitness evidence about whether there actually is a hyperdimensional "war" for control of the third dimension of our solar system. For example, who is the "aggressor" or "invading party" in this hyperdimensional "war"?

First, Earthling humans may be the aggressor. The secret US facility on Mars that itself uses jump-room technologies may be perceived by the hyperdimensionals as an aggression by Earthling humans, inviting counter-attack by the hyperdimensional "enemy" at the time-travel jump gates of the Mars base.

Second, the hyperdimensional "enemy" may be the aggressor. The US facility on Mars may be a strategic forward position by the Earthling human forces in alliance with the Martian humans (living under the surface of Mars) to thwart off a threatened occupation of the Mars third time-space dimension by the hyperdimensional "enemy."

Third, the "enemy" may be part of a False Flag operation. The "armed attacks" by the hyperdimensional "enemy" on the US base on Mars may be part of an elaborate False Flag deception scenario to create a faux "war" against the "evil extraterrestrials."

Of these alternatives, it seems most reasonable to conclude that the "hyperdimensional war" around US facilities on Mars as reported by Michael Relfe is accurate and is an example of hostile, competing interaction in the dimensional ecology by intelligent civilizations based in other dimensions.

Archons: Hyperdimensional Controllers in the Dimensional Ecology

Archons are hyperdimensional hidden controllers in the dimensional ecology of our universe. According to researcher Laura Magdalene Eisenhower, archons are hidden negative controllers of humankind, inorganic interdimensional entities that must now be exposed and exorcised from the individual human mind, from our human species, from the planet, and from our universe as a whole, as part of our collective evolution to a new state of consciousness and being.

The ancient Gnostic texts from Egypt, the Nag Hammadi library, describe two types of demonic alien beings that invaded Earth long ago which they call the archons. The first type of archon looks like a reptile, the other like a human embryo, with the same shape and appearance as the "sky fish" photos.[22]

According to Robert Stanley,

> It is time to expose the covert controllers of mankind. I assure you this is not speculation, a hoax, or the figment of peoples' imagination. These parasitic creatures are real and they need to be dealt with immediately so mankind can evolve to the next level of existence.
>
> Although these parasites are not human, they feed off the negative energy / emotions of humans. It is unclear when these cosmic, amoeba-like creatures first came to Earth, but we know they were discovered by shamans in altered states of consciousness long ago and have recently been photographed. The reason everyone is not seeing them on a daily basis is because the creature's energy signature is beyond our normal, narrow range of vision within the electromagnetic spectrum, what scientists call "visible light."[23]

---

[22] See photographs at Webre, op.cit. http://exm.nr/HfQpQj; see also http://bit.ly/1crsAU1

[23] Alfred Lambremont Webre, "Archons - Exorcising hidden controllers with Robert Stanley and Laura Eisenhower", http://exm.nr/HfQpQj

ARCHONS ARE INTRAPSYCHIC MIND PARASITES

In discussing archons, John Lash[24] writes,

> Although archons do exist physically, the real danger they pose to humanity is not invasion of the planet but invasion of the mind.
>
> The archons are intrapsychic mind-parasites who access human consciousness through telepathy and simulation. They infect our imagination and use the power of make-believe for deception and confusion. Their pleasure is in deceit for its own sake, without a particular aim or purpose. They are robotic in nature, incapable of independent thought or choice, and have no particular agenda, except to live vicariously through human beings. They are bizarrely able to pretend an effect on humans, which they do not really have.
>
> For instance, they cannot access human genetics, but they can pretend to do so, in such a way that humans fall for the pretended act, as if staged events were taken for real. In this respect, archons are the ultimate hoaxers. This is the essence of "archontic intrusion," as I call it. The trick is, if humanity falls under the illusion of superhuman power, it becomes as good as real, a self-fulfilling delusion.[25]
>
> To sort out and clarify what the Sophianic narrative may have to say about the test of the archons is a great challenge to our understanding of the Gnostic message and how it can benefit humanity today.[26]

Researcher Robert Stanley maintains that humanity must now take a scientific approach to identifying archons and exterminating them in the human dimensional ecology. There is *prima facie* documentary (photographic) evidence substantiating the existence of archons.[27]

---

[24] John Lash, "Who Wrote the Reptilian Agenda?" http://bit.ly/1cLWRXT
[25] Robinson et al., NHC II, 1:28.16.
[26] Webre, "Archons - Exorcising hidden controllers with Robert Stanley and Laura Eisenhower," http://bit.ly/1crsAU1.
[27] See photographs at Webre, op.cit. http://exm.nr/HfQpQj; see also http://bit.ly/1crsAU1.

Mary Rodwell: Hyperdimensional Intelligent Civilizations Interact with Earthling Humans in a Multiverse

The dimensional ecology of the Omniverse hypothesis is supported by replicable empirical evidence gathered by the Australian Close Encounter Research Network (ACERN) and researcher Mary Rodwell, who has developed an evidence-based model of how hyperdimensional intelligent civilizations interact with Earthling humanity in the Omniverse. Her model demonstrates how hyperdimensional civilizations employ the dimensional ecology to communicate ethically and cooperatively with humans at an individual and collective level.[28]

In her article "Extraterrestrials, human consciousness, and dimensions of soul: The intimate connection," Mary Rodwell presents her research into more than 1,800 cases of abduction by teleportation of contactees by hyperdimensional civilizations in the dimensional ecology. She concludes that the hyperdimensional extraterrestrials, the human subjects, and their abductions potentially demonstrate "a new reality: a multiverse."[29] She writes,

> The human connection to extraterrestrial intelligences: does it transcend space and time? I believe we are trying to explore a multiverse through the eye of a needle.
>
> The majority of the public and most researchers believe that Contact or encounters with extraterrestrials are defined through two, very limited windows of Contact experience. Firstly, the Abduction scenario, when the individual may feel paralyzed, and taken in a trance, a deep sleep-like state on-board craft. Secondly, the contactee, who has full conscious recall of their Encounter. My research, however, suggests a third, far broader category, which is far more common and less physically intrusive. It encompasses some of the

---

[28] Australian Close Encounter Research Network (ACERN). http://bit.ly/18DVPed.

[29] Rodwell, "Extraterrestrials, human consciousness, and dimensions of soul: The intimate connection," http://bit.ly/19TrVWY.

above, but embodies more subtle patterns of experience. If we broaden the mandate to be inclusive of this third category, we stand to gain a far more fascinating picture of Contact and how or why these intelligences are interacting with us.

In this article I explore all forms of Encounters and ask why research with Experiencers of Contact with these intelligences offers such diverse and conflicting data. What approach is the best for the Experiencer seeking answers and verification of [such] encounters? I explore the benefits of the holistic approach to gathering data.

[The article sets out] misconceptions regarding data revealed by hypnosis, and [questions if this] extraordinary data is confabulation, fantasy, or potentially demonstrating a new reality: a multiverse.

The testimony from Experiencers challenges not only consensus reality, but our dearly held religious and spiritual beliefs, and demonstrates an intimate and complex association with extraterrestrials, a connection that spans both space and time, including the realms and dimensions of soul.[30]

Midway through her article, Mary Rodwell goes on to explore the multiverse as well as the data supporting a dimensional ecology between various dimensions in this multiverse. She writes,

> This phenomenon asks us to be courageous and challenge consensus reality, to be "wise" enough to accept what we don't know, cognizant that our present understanding of reality is being framed by a science in transition. Concepts such as quantum reality, the holographic universe, unity consciousness, [are] theories which suggest that everything is interconnected through consciousness. This is very similar concept to what those with Encounters convey, but experientially. Their experiences demonstrate this alternate, multifaceted paradigm that could prove to be the "true nature of reality." I feel that we as a species are being challenged by such Contact to relinquish our old reality programs. Encounters [are] a modern-day "shamanic experience," designed to break down outmoded programs of reality and allows for us to become fully conscious and operate in a multidimensional reality. In *Awakening* I

---

[30] Ibid.

discuss this as the "realization event," when the person finally accepts that the old paradigm no longer works as they embrace the new one. It can be terrifying to have all your beliefs of reality challenged and to accept an expanded reality. But like the apprentice shaman, they face their ultimate fears, to explore and cope with a multidimensional perspective of reality. The shamanistic experience was found to be parallel in outcome to Encounters, bringing transformative, heightened psychic and healing abilities to the individual.[31] ...

But, if true, [these data] indicate interactions/encounters not only in physical form, but when we are separated from our physical body, such as in an astral or out of body (OBE) soul state. But added to this seem to be interactions through many lifetimes, and the in-between-lives and before our present human incarnation, encounters that continue after our present physical life ceases. So to all intents and purpose the interactions are in both physical and nonphysical [forms] through many timelines, as we are educated and assisted in physical realms and the realms and dimensions of soul.

This information is profound and answers ... one of the most pressing of questions asked by the Experiencer, especially [those who] have felt victimized by their encounters. To their astonishment, when the question regarding consent is asked, they may be shocked to discover this permission was given, before their present human incarnation. They describe themselves as an intelligence, which appears to them to look like a bright orb of light, their soul essence. They vocalize: "I agreed, but it was before I came here." Some add, "I have a job to do. To assist in the evolution in human consciousness." Hundreds have verbalized such statements.

I recall sharing this information, with one well-known author and fellow researcher. He totally dismissed the data, and argued this information is a form of delusion, a program instilled by these Intelligences hoping to trick the individual into accepting their encounters. I find that explanation very difficult to believe, given how my clients resonate at some deep level, the fact that it makes so much sense to them. I question who is paranoid. Do we trust or negate what resonates to the Experiencer, or what we as observers prefer to believe? I do not have the answers, I do not pretend I have, but I do

---

[31] *Ibid.*, citing Wilson (2000).

have concerns when anyone articulates such a mandate, without any data to prove such a negative and fear-based perspective, [it] is unhelpful to this research.[32]

Ms. Rodwell quotes hyperdimensional experiencer Tracey Taylor, of Western Australia,

> As far as my own contact experiences I became very compassionate towards the ET beings as there is much humans can learn and benefit by interacting with them. I was fearful of my experiences only when I realized that I seemed to be the only one having them. I didn't know what was going on or why. But there has always been an "equal" exchange. I helped them achieve their genetic goals and in return my healing and psychic ability, plus my life on earth and so on has been enhanced. They also protect me when I ask them to. Basically at the start I misunderstood the Greys. The knowing "ET" part of me initially made a decision to assist them and understanding of this can be overshadowed by fear which stems from our limited human perceptions and reactions to these experiences.[33]

DIMENSIONAL ECOLOGY: THE SOUL DOMAIN

About the "soul domain," Mary Rodwell writes,

> If we are open to what we don't know, it leads us into some fascinating hypotheses. [Consider] the OBE encounter, where many individuals perceive themselves as "orbs" or spheres of light. They describe leaving their body as an "orb of light," in metaphysical terms called a *merkabah*. This happened to Experiencer Peter of Sydney, Australia, who stated, "A beautiful space, I love it, it is my home, we are everything, I'm every shape, I was a born soul, just beautiful, with energy, the energy of the soul, we create, you create, I'm liquid, I'm fluid, I can go here or there, like a bubble, this is the form of soul."...
>
> The "orb" of light is suggestive of the "soul state" when traveling out of body. In some encounters, people relate that their partner cannot be woken up, no matter how they called out to them, shouted, or

---

[32] Ibid.
[33] Ibid.

shook them. In regression hypnosis, I had one explanation. The "awake" person perceived an orb of light entering the bedroom a bright, greenish "orb." It moved about the room, and it was so strange and scared the person. Very soon after it appeared, the sleeping person awoke. In regression this orb of light proved to be the astral body or OBE "soul state" of the sleeping person, returning to [the] physical body.

In another instance, it was demonstrated that we may inhabit other "life-forms" while in this OBE state. An Experiencer, one of what I call a Star seed, felt she had "dual consciousness," articulated as part human, part ET. In regression, she described being on-board a craft, saw herself in the form of a Grey being, working as a scientist. I asked where her human form was located. She answered, it was on the craft but inanimate, while her consciousness inhabited her "ET self." I asked how she returned to her human body. She described a ball of light leaving her extraterrestrial form to re-animate her human body. She said it was like she changed one biological overcoat for another. Many other individuals have experienced this swapping of physical form to an extraterrestrial form when in an OBE state.

This research offers an insight that demonstrates an extraordinary perspective of our intimate connections to these intelligences, and invites more questions as to the true nature of reality. It examines concepts such as interspecies genetic links, individuals told they may have DNA from one or more extraterrestrial life forms, links that transcend space-time, and interactions with souls before and after human incarnation. Some extraordinary data came from a 15-year-old boy I worked with in the UK in 2008, when on a lecture tour. The data was intriguing, especially given the age of the young man. There were many fascinating aspects to the case, and I did hypnotic regressions with two members of a family of four. They had a joint encounter and experienced "missing time." I regressed the grandmother, which was extremely successful in dealing with her long-held fears, fears she finally released through this process. Her 15-year-old grandson I will name John (not his real name). John was very uncertain he wanted to have hypnosis, fearful of what he would discover. However, the positive, healing outcome for his grandmother encouraged him to proceed. In the regression he spoke about the profound nature of what he termed "new souls." We have to be reminded this was a 15-year-old boy speaking about concepts he had no awareness of consciously. In one encounter with the Greys, he also recognized his departed relatives, describing them appearing as "orbs" of light. He said they were there to support him, to stop him being frightened.[34]

It is certainly appropriate for us to ask why the Earth's time-space dimension is so important to the hyperdimensional intelligences. Mary Rodwell concludes,

> A complex extraterrestrial program, genetic links, multilevel educational programs, coupled with soul interaction, is not congruent with a subversive agenda. Even to my limited human logic, that is nonsensical. Extraterrestrials have technology to take us out in a nano-second if that was their desire. No, these encounters suggest a far different hypothesis, something far more profound, and the reward is transformation of consciousness.[35]

---

[34] Ibid.
[35] Ibid.

Chapter 5
Intelligent Civilizations' Governance Authorities

The Omniverse is, by the evidence, highly populated. The multiplicities of intelligent civilizations in the universes of the multiverse are organized via governance authorities. Please recall that a highly conservative estimate of the number of intelligent civilizations in our universe alone is 100 billion, and that Stanford physicists Linde and Vanchurin have calculated that there are

10 raised to the power of $10^{10,000,000}$

physical universes in the Multiverse. Even this estimate assumes that there are only twelve intelligent civilizations in our galaxy, whereas we know now there are at least 50 million Earth-like planets.[1]

The Anglo-USA official governmental and academic culture at present assumes that Earthly institutions and actors are the

---

[1] Linde and Vanchurin, "How many universes are in the multiverse?" http://bit.ly/HtbdEr. Krum, "How many alien civilizations are there in the galaxy?" http://bit.ly/16ygWlH.

only stakeholders in outer space.[2] The database of *prima facie* evidence for intelligent civilizations' governance authorities is in its early stages of development and will improve as disclosure of official extraterrestrial monitoring and liaison programs by governmental military and intelligence agencies, such as those of Russia, goes forward, and as information that Earthling governments have developed is released into the public domain. Moreover, once open public contact and interaction with intelligent civilizations commences officially, these civilizations will share more information about their governance authorities and governance in our larger universe, as well as in parallel universes of the Multiverse.[3]

NATURE OF THE EVIDENCE FOR EXISTENCE OF INTELLIGENT CIVILIZATIONS' GOVERNANCE AUTHORITIES

As of this writing, there are at least three sources in the public domain of *prima facie* Exopolitical evidence for legally constituted intelligent civilizations' governance authorities that exist in the dimensional ecology of the multiverse and have jurisdiction over a defined dimensional territory, such as the Milky Way Galaxy. This *prima facie* evidence includes types of eyewitness and documentary evidence that are within the protocols of evidence acceptable to the science of Exopolitics.

Source #1: *Replicable scientific remote viewing*

Scientific remote viewing is one source of replicable empirical evidence for intelligent civilizations' governance authorities within

---

[2] Newlove-Eriksson and Eriksson, Johan, "Governance beyond the global: Who controls the extraterrestrial?" http://bit.ly/1g6woaX; Cockell, "Liberty and the limits to the extraterrestrial state," http://bit.ly/1crfovd.

[3] See, for example, Walia, "Russian Prime Minister confirms the existence of intelligent extraterrestrial ;ife," http://bit.ly/1hYqkT4; and Hellyer (a former Canadian Defense Minister), "Extraterrestrials want to help mankind," http://bit.ly/1hYqxWn.

the dimensional ecology of the multiverse. Empirical evidence derived from replicable communication with representatives of governance authorities of our Milky Way Galaxy, using standard laboratory remote-viewing protocols, has been reported to the public.[4]

According to the Farsight Institute, remote viewing is

> a controlled and trainable mental process involving psi (or psychic ability). Remote-viewing procedures were originally developed in laboratories funded by the United States military and intelligence services and used for espionage purposes. The scientific understanding of the remote-viewing phenomenon has greatly advanced in recent years, and as a result the process of remote viewing can now be reliably demonstrated in both laboratory and operational settings. There are a number of styles of remote-viewing procedures that are popularly practiced, such as Scientific Remote Viewing (SRV), Controlled Remote Viewing (CRV), Hawaii Remote Viewers' Guild procedures (HRVG), as well as a few others. Remote viewers use one or more of these styles to gather descriptive data of a "target," which is usually some place or event at some point in time.
>
> Remote viewing is always done under blind conditions, which means that the remote viewer must know nothing about the target when conducting the remote-viewing session. All of the various styles of remote viewing require both training and regular practice in order for a remote viewer to become proficient. Remote viewing is normally considered a controlled shifting of awareness that is performed in the normal waking state of consciousness, and it does not typically involve an out-of-body experience, hypnosis, an altered state of consciousness, or channeling.[5]

Remote viewing is a consciousness technology involving a standard laboratory protocol that can access intelligences in nonlocal subspace in the multiverse and produce replicable communication. Remote viewing thus provides a means to transmit and receive scientifically replicable information between a trained individual based on Earth in our time-space hologram

---

[4] Brown, Cosmic Explorers, pp. 147-177.
[5] Farsight Institute, "Remote Viewing," http://bit.ly/1iB4cNb.

and a representative of an intelligent civilizations' governance authority in some adjacent dimension of our universe or parallel universes in the Multiverse. Remote viewing is thus an ideal laboratory protocol and tool for support and verification of the dimensional ecology of the Omniverse hypothesis.

Replicable data from an experiment by scientific remote viewers associated with the Farsight Institute suggest that a spiritually and technologically advanced galactic federation of worlds exists. It can be described as a sort of loosely organized dimensional government of our Milky Way Galaxy.

For example, the following is a transcript of a scientific remote viewing session with a Milky Way Galactic federation leader:

> *Remote Viewer*: As soon as I went into his mind [that of a galactic federation leader], I re-emerged in space. That is where I am now. I am outside of the Milky Way, looking onto it. Dotted lines have been drawn over the image, dividing up the galaxy-like quadrants.
>
> I am being told that there is a need for help. They need us. I am getting the sense that they need us in a galactic sense, but I seem to be resisting this. They are so much more powerful than humans; it just does not make sense why they would need us.
>
> The leader is sensing my resistance and redirecting me to a planet. OK, I can see it is Earth. I am being told that there will be a movement off the planet in the future for humans. I am just translating the gestalts now into words. But the sense clearly is that Earth humans are violent and troublesome currently. They need shaping before a later merger. Definitely humans need to undergo some sort of change before extending far off the planet.
>
> *Monitor*: Ask if there are any practical suggestions as to how we can help.
>
> *Remote Viewer*: I am being told in no uncertain terms that I am to complete this book project. Others will play their parts. There are many involved, many species representatives, groups.
>
> *Monitor*: As who else we should meet using remote viewing, or another technique?

Remote Viewer: Only the Martians. Hmmm. I am being told that our near-term contact with extraterrestrials will be limited to the Martians for now, at least in the near future.

Monitor: Ask if there is new information that we need to know but do not know.

Remote Viewer: This fellow is very patient. He knows this is hard for me. He is telling me that many problems are coming. There definitely will be a planetary disaster, or perhaps I should say *disasters*. There will be political chaos, turbulence, an unraveling of the current political order. As we are currently, we are unable to cope with new realities. He is telling me very directly that consciousness must become a focal concern of humans in order for us to proceed further.

He is right now tapping into *your* (my monitor's) mind. It is like he is locating you, and perhaps measuring or doing something. He is telling me that you are very important in all of this. We must come back here—their world—at later dates. We will be the initial representatives of humans as determined by consciousness. He is telling me that consciousness determined our arrival point. There is more. We are not saviors, just initial representatives. He wants me to get this straight.

I am getting the sense that he wants us to understand that we have a responsibility to represent fairly. This is not to go to our heads. This is just our job now, and **we** all have jobs. He is also telling me that I am doing a fairly good job at writing this down.

He likes your sense of humor. He says that there will be *lots* of activity in the future, of the day-to-day sort. But for now, we are to focus on the book. The book is important, and they will use it.[6]

Source #2: Contactee, telepathic, and documentary evidence

There also exists much *prima facie* witness evidence for intelligent civilizations' governance authorities, consisting of eyewitness contactee and telepathic interaction with representatives and intermediaries of reported intelligent civilizations' governance authorities.

---

[6] Brown, Cosmic Voyage, p. 92.

The details contained in this evidence vary widely and are not necessarily consistent with each other or with known history. More than anything, this tends to be *prima facie* evidence that such authorities do exist, rather than evidence of the content of their mission, role, and history. Wherever possible, the contents of the *prima facie* evidence about governance authorities are checked for congruency with comparable information.

Role of Governance Authorities in the Origin of the Human Race

Reported telepathic communications with representatives of regional galactic governance councils in the dimensional ecology of our galaxy have provided *prima facie* evidence of the possible origins of the Earthling human race that is congruent with scientifically derived information about human DNA.

For example, one telepath who is in telepathic contact with a reported regional galactic governance council states that a consortium of advanced intelligent upper-dimensional extraterrestrial and interdimensional civilizations originally developed *Homo sapiens* as an intelligent being with twelve-strand DNA, as a species that was to have been a guardian of the third dimension of time-space in our dimensional ecology on Earth.

In discussing the development of *Homo sapiens* by a consortium or council of advanced, ethical, intelligent hyperdimensional civilizations, including Pleiadians and Sirians, telepath Patricia Cori states,

> Intensive studies were made of the existing environments of Earth—the distinctive plant and animal kingdoms—and detailed investigations were conducted into how biodiversity resulted as a reflection of various geographical and climatic variables. It was discerned that such diversity would provide ideal conditions for the seeding of extraterrestrial species, as their original environments could, in many ways, be replicated in the ecosystems of Earth.

Unexplained in your archeological and missing-link evolutionist theorist is the isolated emergence upon your planet of four distinct seed races. These are archetypes of master races whose fundamental genetic material formed the primary "substance" of your race, while the vibrational patterns and sequencing of extradimensional beings (those of higher realms) were woven into the intricate light codes of your incredible twelve-stranded DNA. The genetic material of these four primary races [black, red, yellow, and white] was united into the blueprint of the species of Homo sapiens.[7]

MULTIPLE REPORTS OF A REGIONAL GALACTIC GOVERNANCE COUNCIL

Patricia Cori's telepathic communications with a reported regional galactic governance council in the dimensional ecology of our galaxy is congruent with recent empirical data and exopolitical analysis indicating that there is a similar regional galactic governance council, concerned with Earth's ecology and humanity, to which many of the same reported planetary civilizations belong. There is also *prima facie* documentary evidence of intelligent civilizations' governance authorities, consisting of photographs and video recordings of decloaking by spacecraft of these authorities over major Earth cities on predicted dates.

Former Canadian NORAD officer Stanley A. Fulham has reported interacting with representatives of a regional galactic governance council whose membership includes the upperdimensional, ethical Pleiadian and Sirian civilizations, who are concerned with the development of life-bearing planets in this area of our galaxy, including Earth.[8] It is important to know that

---

[7] Webre, "Author: ET council seeded Homo sapiens as intelligent beings with 12-strand DNA," http://exm.nr/1h6KiOF.

[8] I have discussed this evidence in many articles, especially the following:
"ET council: We will increase UFOs, address U.N. in 2014, renew ecology in 2015," http://exm.nr/HtiNyQ; "More predicted UFO sightings over New York confirm ET will intervene in ecology," http://exm.nr/17CmVq5; "Stanley Fulham dies, warned ETs will intervene & save Earth's collapsing ecology," ttp://exm.nr/16T0j6m.

Fulham was a life-long NORAD officer who directed the NORAD radar that tracked hyperdimensional ET UFOs across and around the North American continent.

When I interviewed Fulham, he stated that a regional galactic governance authority decided in January 2010 to put aside the law of nonintervention. At a solemn meeting, the regional authority decided to intervene with their technology to clean Earth's atmosphere before an environmental collapse occurs on Earth, as has happened on many other inhabited planets with civilizations similar to our own. The authority did so, according to Fulham's information, after concluding that our human technology could not prevent an environmental collapse and species extinction on Earth.

According to Fulham, this is a rare decision by the authority, resulting partly because authority members wish to preserve the "unique positive qualities of our human population." The authority, according to Fulham's information, "consists of the intelligent civilizations of the Pleiades, Orion, Sirius, Bootes, Alpha Centauri, Comsuli, Zeta Reticuli, and Pouseti."[9]

The regional galactic governance council, according to Fulham, has had a caretaker role for our planet for about the last million years, and has effectively maintained Earth under protective quarantine following an attempted invasion and takeover of Earth.

STANLEY A. FULHAM'S DECEMBER 3, 2010, STATEMENT

Just prior to his untimely death, Fulham published the following statement on his website:

> I had advised my readers and UFOlogist supporters I would notify them by e-mail on the next major UFO display in their "intervention process" with mankind which started in their major display over New

---

[9] Webre, "ET council: We will increase UFOs,"http://exm.nr/HtiNyQ.

York on 13th October 2010. The prediction, documented in my book *Challenges of Change* ... stated that "there would be a major display over our principal cities" on that date.

There was a great deal of speculation about this change in the Transcendors prediction. ["Transcendors" are Fulham's interdimensional telepathic connection to the regional galactic governance authority] In late September 2010 I asked the Transcendors about this significant change. They replied: "The Pleiadians are much more aware of the consciousness of mankind than you are—after all, they have been observing you for millions of years. They came to the conclusion that mankind (generally speaking) is not prepared for a massive display of their UFOs over our principal cities. They would regard it as a hostile intervention—a prelude to invasion. Accordingly, the Alien Council decided about three-weeks before their 'display' they would focus on one of our principal cities, New York. It is recognized worldwide as our most cosmopolitan city with a population not likely to be overwhelmed by the sight of UFOs; however, the 'display' date of 13 October 2010 would remain. There were other major UFO sightings on that date all over the world. But the Transcendors caution that they were not part of the Council's principal focus over New York. The New York display is simply the beginning of the Aliens' "intervention process."

In my last "Reading" with the Transcendors on Tuesday, October 13th, 2010, the Transcendors advised me of the following predictions on the Pleiadian UFO displays:

A. There will be a major UFO display over Moscow between the end of the first part of January 2011 and the second week of January 2011.

B. This "display" will be followed by a major display over London approximately seven days later.

C. Interventions will then accelerate, not so much over our cities, but dispersed over our continents, with sightings increasing in duration. The intent of these interventions is to increase mankind's acceptance of the alien phenomena, so that, hopefully, we will be prepared to accept a face-to-face encounter and communicate, perhaps as early as next year (2011).

D. Objective—a dramatic introduction of the alien reality—an appearance at the United Nations.

The Alien "intervention process" will provide a philosophical challenge to mankind as nothing else has ever done. It will challenge our most fundamental beliefs. Did Christ teach to this extraterrestrial world? Are they aware of Allah and his hatred for the heathens and infidels (which, of course, must include the aliens—they also have souls!)? Do they have any religions? —Or spiritualism?

The UFO "intervention process" will have a major impact on the US and Russian governments [and] will be the catalyst that will break the conspiracy of silence."

UFOlogists can then divert their focus to the challenges the new alien technologies will bring to mankind.

Due to my declining health, this may be my last writing on this subject. In the meantime, I wish all my readers and UFOlogist supporters the very best in their future work.

Stanley A. Fulham, December 3rd, 2010

October 13, 2010: Decloaking of Authority's Spacecraft

The original plan of the regional galactic governance authority, taken at its meeting in January 2010 (Earth time), had been for simultaneous decloaking of the authority's spacecraft on Oct 13, 2010, over major world cities. The purpose of these appearances was to acclimatize Earth humans to the authority's presence and decision to intervene.

Fulham stated in his Exopolitics radio interview that the regional council had scaled this plan down to one city and had chosen New York City for an initial Oct 13, 2010, spacecraft "decloaking" because New York City is a global, cosmopolitan city with a blasé population that would not be frightened of their appearance. There are multiple, independent evidentiary sources that prove the October 13, 2010, UFO sightings over New York

City revealed intervention by a non-Earthly intelligent civilization, and were not the result of other causes, such as released balloons.[10]

Synchronistically, on October 14, 2010, a day after the predicted October 13, 2013, sightings over New York City, Dr. Mazian Othman, director of the UN Outer Space Office, delivered a wide-ranging 28-minute video press conference at the United Nations in New York, during which she stated that "extraterrestrial life is a possibility" and remarked that the United Nations must ready itself for extraterrestrial contact. Following the October 13, 2013, sightings over New York City, the regional galactic governance authority source predicted a series of sightings of their spacecraft over Moscow, Russia, and London, UK.

WERE THE UFOS IN RUSSIAN SKIES IN DECEMBER 2010 FULHAM'S PREDICTED UFO WAVE?

On December 29, 2010, I reported the following:

> Russian Television (RT) is now reporting UFOs beginning to appear over the skies of Russia and has released a video and news article confirming these sighting. Former NORAD officer Stanley A. Fulham released a statement just prior to his recent death from pancreatic cancer that ... a regional galactic governance council would begin more frequent UFO appearances in 2011, starting with "a major UFO display over Moscow between the end of the first part of January 2011 and the second week of January 2011."
>
> Query: Is this wave of UFO sightings now beginning over Russia part of the UFO wave predicted by Stanley A. Fulham in his final statement?[11]

One Russian report stated,

> In early December UFOs began to frequent the skies of the capital of the republic of Kalmykia, located in southern Russia. Every ten days

---

[10] Ibid.

[11] Webre, "Are the UFOs appearing in Russia skies in Dec. 2010 Fulham's predicted UFO wave?" http://exm.nr/17y2DQW. The video embedded in this article shows the Russian television report as transcribed below.

the residents of Elista have seen UFOs of unknown nature in different parts of the city.

Latest information on this phenomenon appeared on 22 December, 2010, when witnesses observed this type of flying objects in different parts of Elista. So some citizens say they have seen in the sky two concentric circles spinning in opposite directions. Others speak of a triangular object radiating light.

Scientists cannot give a rational explanation for these phenomena. For example, Professor of Theoretical Physics at the University of Kalmykia, Mijaliáyev Badme, does not rule out the possibility of the existence of an alien civilization, but also sees no clear evidence of the existence in the appearance of UFOs in Elista sky. Meanwhile, the ufologists say that UFOs appear in other parts of Russia.

EXTRATERRESTRIAL INTERVENTION, ACCLIMATIZATION, AND UFO WAVE PHENOMENA

The December 2010 UFO wave phenomenon in Russia does seem related to the extraterrestrial acclimatization process set out in dimensional communications from the regional galactic governance council, as published by Stanley A. Fulham in his book *Challenges of Change*[12] and in his December 3, 2010, statement. Two paragraphs in that statement seem particularly relevant in this regard:

> The UFO "intervention process" will have a major impact on the US and Russian governments. This will be the catalyst that will break the "conspiracy of silence."

and

> Interventions will then accelerate, not so much over our cities, but dispersed over our continents with sightings increasing in duration. The intent of these interventions is to increase mankind's acceptance of the alien phenomena, so that hopefully, we will be prepared to

---

[12] Fulham's Challenges of Change is the product of ten years of research and interaction with a council-connected source, and contains a blueprint of the council plan for 2010 to 2015.

accept a face-to-face encounter and communicate, perhaps as early as next year (2011).

It seems reasonable that the December 2010 UFO wave over Russia—if the Russian government is a prime target of the regional galactic governance authority in breaking "'the conspiracy of silence" by Russia and the US over the extraterrestrial presence, and its intent to save the Earth's ecology—is part of the new UFO waves predicted by Fulham prior to his untimely death from pancreatic cancer. Independent analysis has concluded that Fulham's pancreatic cancer may have been accelerated by covert human agencies in order to silence him on the issues of the regional galactic governance council.

Independent Confirmation of Hyperdimensional Council

On January 13, 2011, I reported the following:[13]

> A remarkable UFO wave over Moscow, Russia, in early January 2011 has reportedly confirmed the predictions of an apparent regional galactic governance council that its interdimensional spacecraft would appear over Moscow, Russia, at that time. On December 3, 2010, Stanley A. Fulham had predicted a major UFO display over Moscow early in January 2011, with another such display over London approximately seven days later.
>
> This appearance of apparent Pleiadian UFO craft over Moscow, Russia, marks the second independent occasion on which interdimensional messages inspired by an apparent galactic governance council and predicting an overflight of Pleiadian spacecraft over a major urban area have in fact been accurately fulfilled.
>
> The first occasion was a council prediction through Fulham that galactic governance council ships would appear over New York on

---

[13] Webre, "January 2011 UFO wave over Moscow is 3rd independent confirmation of ET council," http://exm.nr/1iB7osb.

the night of October 13, 2010. In fact, UFOs did appear then and were filmed.[14]

The second occasion was the January 2011 UFO sightings over Moscow, Russia. These documented sightings of apparent Pleiadian UFO galactic governance council ships fulfilled Fulham's December 3, 2010, prediction.

The Moscow UFO Sightings and the US-Russia "Conspiracy of Silence"

The January 2011 UFO sightings over Moscow underscore an apparent determination by the hyperdimensional council to break the US-Russia "conspiracy of silence" around the extraterrestrial presence and the dire danger that the Earth's ecology and our Earthling human survival may be in, should this information be correct.

In his last communiqué on December 3, 2010, before his untimely death, Fulham emphasized that breaking the conspiracy of silence by the US and Russia on the extraterrestrial presence was a major goal of the galactic governance council.

London or Paris UFO Sightings in Mid-January 2011?

According to Fulham's December 3, 2010, communiqué, the next wave of Pleiadian UFO overflight displays would be followed by a major display over London approximately seven days later. A close associate of the now deceased Fulham sent me the following advisory on January 12, 2011.

> Stan mentioned he wasn't exactly sure for the longest time if the next event was going to be London or Paris, but after much deliberation,

---

[14] To see videos and images of these spacecraft embedded in my articles, see Webre, "More predicted UFO sightings over New York confirm ET will intervene in ecology," http://exm.nr/17CmVq5, and Webre, "January 2011 UFO wave over Moscow is 3rd independent confirmation of ET council," http://exm.nr/1iB7osb..

finally settled on London. He was hoping to have another reading with the Transcendors to try to "fine-tune" a more exact date and place, but was too weak and ill. I thought I should mention that to someone in advance, just in case it turns out to be Paris, and the time frame is a bit off.

Asking the Key Question

We need to ask whether the "Galactic Governance Council," as described by Fulham and as apparently confirmed by the two independent UFO sightings, might in fact be instead a "False Flag" operation, a Psyops, disinformation, or random UFO sightings—or is it what it appears to be, an intervention by a galactic governance authority?

That is a fair question. It has been raised and continues to be raised about the UFO sightings on October 13, 2010, and on November 24, 2010, over New York, and in January 2011 over Moscow.

Let us examine the various key aspects of the galactic governance authority intervention hypothesis to date.

1   *Technology of hyperdimensional ET False Flag operations*

There are multiple ongoing studies of planned hyperdimensional ET False Flag operations. I have listed the various psyops technologies necessary for an extraterrestrial False Flag operation, in the style of another 9/11.[15]

In the absence of any other factors, and taking each of the October, November, and January sightings, a case could be made that each of these may have been either random UFO sightings or a hyperdimensional ET False Flag operation, not part of an intentional, interdimensional intervention by an intelligent

---

[15] Webre, "Whistleblower exposes attempted ET manipulation, false flag at 'Festival'," http://exm.nr/17zzQLM.

civilization that so announced itself beforehand. But there are other factors to consider, notably the professional expertise of Fulham regarding extraterrestrial UFOs, given that he had directed the NORAD radar that tracked hyperdimensional ET UFOS across and around the North American continent.

2   Fulham's interdimensional communication with the "galactic governance council"

Stanley Fulham was exceptionally capable in recognizing and tracking extraterrestrial UFO craft. As a lifelong NORAD officer, he recounts having routinely tracked extraterrestrial UFOs over North America. From the perspective of a hypothetical regional galactic governance council, Fulham was probably an ideal contact, because he held the highest professional credentials in the recognition of extraterrestrial UFO craft. He spent ten years in study and interdimensional communication around the issue of the regional galactic governance council and its members, functions, and operations. That knowledge was passed onto us in a substantial, serious book.

3   Timeline of possible galactic governance council intervention

The apparent regional galactic governance council's indication of 2014 and 2015 as years for a potential intelligent civilization decloaking and intervention in Earthly affairs is especially significant because of the four "Tetra" lunar eclipses separated by two solar eclipses scheduled for 2014-15, an astronomical configuration that last occurred in 1492, the year Columbus discovered America, and will not occur again for another 1000 years.[16]

Additionally, one interpretation by modern hermeneutics, the analysis of prophecy, views the Tetra lunar eclipses of 2014-2015

---

[16] "The 4 lunar eclipses in 2014-2015 (Tetra) all on Jewish feast days, and the last one for 21st century is significant," http://bit.ly/1ctIzyU.

as the prelude to a possible interdimensional landing on Earth of positive intelligent civilizations in a "Messianic" support of human society. This view is congruent with the regional galactic governance council source regarding extraterrestrial intervention in 2014 and 2015, fulfilling a supportive role in the dimensional ecology of Earth.[17]

What if the extraterrestrial interventions predicted by the regional galactic governance council source to occur in 2014 and 2015 do not happen? If they do not, that lack will not detract, in my opinion, from the *prima facie* evidence suggesting the existence of galactic governance councils, because the promulgation of intentional misinformation or disinformation appears to be a deliberate strategy of these councils when dealing with the hostile environment of terrestrial governments who may want to attack any decloaked spacecraft of the governance council.

Evaluating the Legacy of Stanley A. Fulham

Stanley Fulham's legacy, taken at this most literal level, is potentially of a well-documented revelation of the ecological intervention planned by a regional galactic governance council, and of three separate fulfilled predictions of UFO decloakings of interdimensional craft of the Pleiadians, the reported "point persons" of this council.

The analysis and evaluation by UFO expert Randy Kitchur demonstrates beyond a reasonable doubt that Fulham's predictions of UFO displays over New York City on October 13, 2010, over Moscow early in January 2011, and over London approximately seven days later were literally fulfilled. The implications of the fulfillment of the Fulham UFO predictions, when read concurrently with the substance of his well-

---

[17] Webre, Interview with Peter Kling, "Part 2 - Peter Kling: Christ returns as an Extraterrestrial with the armies of the multiverse to defeat the NWO?" http://bit.ly/JSviEI.

documented narrative, are substantial. The empirical evidence of the fulfilled Fulham UFO predictions confirms a working hypothesis that a regional galactic governance council in fact exists and is now intervening in the human environment for beneficial ecological reasons.

Some commentators claim that the UFOs in the New York., Moscow, and/or London UFO displays are actually part of a Project Bluebeam holographic display, or a display of terrestrial black-budget UFO craft. This scenario would have it that the Fulham material is actually a covert terrestrial military intelligence "legend" created as part of a False Flag operation to impose a global New World Order in order to supposedly defend Earth against an extraterrestrial invasion or landing. A variant of this argument is the claim that the predicted Fulham UFO displays were by an extraterrestrial civilization that is inimical to the aspirations of humanity for sovereignty and self-determination. However, Kitchur's expert UFO evaluation has demonstrated that the interdimensional UFO craft in these three displays are not part of a terrestrial holographic Project Bluebeam operation nor are they terrestrial anti-gravity UFO craft.

Nor is there any evidence that Fulham was enrolled in a deception operation by a manipulatory extraterrestrial race. Fulham, a deeply spiritual person, as well as a lifelong and highly trained NORAD officer who routinely dealt with UFOs at NORAD, spent ten years in interdimensional communication around the subject matter of the regional galactic governance council and its message about the ecological crisis of Earth. His writings and actions, as well as the actions of the regional galactic governance council in delivering these materials to him, and in creating the separate interventions over the three cities as evidence of their existence and good faith, are *prima facie* ethical acts.

Discrepancies in Interdimensional Information?

Experts in the field of extraterrestrial contactees have commented that contactees of hyperdimensional civilizations are often intentionally given disinformation on certain factual details in order to protect them from attack by human military-intelligence forces on Earth that seek to hide these details. This may well be the case with Stanley Fulham. He appears to have been given accurate information about the dates of the predicted decloakings by the regional galactic governance authority. The rest of the messages of the regional galactic governance authority, about the timing and nature of its intervention in human public affairs, does appear to be disinformation.

Source #3: Information collected by Earthly governments from direct interaction with intelligent civilizations through teleportation and time travel.

Accurate and reliable information collected by Earthly governments from direct interaction with intelligent civilizations through teleportation, time travel, and other space travel missions into our universe and the multiverse will occur upon truthful and complete governmental disclosure.

I have interviewed several whistleblowers from secret US extraterrestrial liaison programs that have involved teleportation and time travel, including former US chrononauts Andrew D. Basiago, William B. Stillings, and Bernard Mendez, former US serviceman Michael Relfe, and others. Research is ongoing into issues of intelligent civilizations governance and will be reported as it is developed.

PART III

Spiritual Dimensions of the Omniverse

CHAPTER 6
THE INTERLIFE DIMENSIONS

Prima facie empirical evidence supports the hypothesis that the Omniverse consists of the totality of parallel universes in the Exopolitical dimensions (the multiverse) plus the spiritual dimensions. The parallel visible universes (or the multiverse) can be more aptly termed the Exopolitical dimensions, since the intelligent civilizations of souls based in the spiritual dimensions of the Omniverse incarnate in the multiverse in order to undergo a variety of life experiences as a diversity of intelligent creatures for the purpose of moral and soul development.

Evidence for the existence of the spiritual dimensions and dimensional ecology between the Exopolitical and spiritual dimensions is provided by the same evidence that reveals the continuation of consciousness in the Interlife after bodily death in the Exopolitical dimensions. The communications of intelligent civilizations of souls based in the Interlife dimensions to Earthlings in the time-space hologram via the dimensional ecology provide the evidence for the existence of souls, the Interlife, and the dimensional ecology itself.

Dimensional Ecology: The Intelligent Civilizations of Souls

By "soul" is meant an individuated, nonlocal, conscious, intelligent entity that is based in the Interlife dimensions. Each one is a holographic fragment of the original Source or Creator of the spiritual dimensions of the Omniverse. The empirical evidence for the existence of souls is derived from a replicable database of more than 7,000 cases of hypnotic regression of soul memories of the Interlife, developed according to a standard protocol. These replicable data report that souls are created as a holographic "egg of Light" drawn from the original Source that is a "Sea of Light."[1]

Instrumental Transcommunication (ITC)

Instrumental Transcommunication (ITC) has developed replicable empirical evidence of interactive communication between living humans in our time-space hologram and the apparent intelligences of deceased humans based in either (a) a hyperdimension of the Exopolitical dimension, or (b) the Interlife in the spiritual dimensions. ITC uses electronic technology, such as video technology, audio technology, computer technology, Internet streaming video or audio, and cell phones.[2]

Discussing the data from the dimensional ecology in the afterlife dimensions, Instrumental Transcommunication (ITC) researcher Victor Zammit writes,

> According to Mark Macy, a leading researcher in this area, throughout the 1990s, the research laboratories in Europe received extended, two-way communication with spirit colleagues. This was almost daily through telephone answering machines, radios, and computer printouts.

---

[1] See the books by Michael Newton in Sources and Resources. These are: Life between Lives: Hypnotherapy for Spiritual Regression; Destiny of Souls; and Journey of Souls.

[2] For sample ITC images from streaming video on Internet and interaction with "deceased" persons in the human "afterlife" dimension, see http://bit.ly/1aFXy9m.

His book *Miracles in the Storm* outlines how scientists working for the International Network for Instrumental Transcommunication (INIT), which he founded in 1995, received from the afterlife:

- pictures of people and places in the afterlife on television that either appeared clearly on the screen and remained for at least several frames, or which built up steadily into a reasonably clear picture over multiple frames;

- text and picture files from people in the afterlife which appeared in computer memory or were planted on disk or similar recordable media;

- text and images of people and places in the afterlife by way of fax messages.[3]

Instrumental Transcommunication (ITC) provides replicable empirical evidence of the dimensional ecology between our time-space hologram in the Exopolitical dimensions and the Interlife dimension, as well as of survival of the soul after death. ITC produces physical trace evidence in the form of voices and images that correspond to specific identifiable former living and now "dead" persons.

ITC AS NONLOCAL COMMUNICATION

Ervin Laszlo writes,

> The authenticity of ITC is not entirely beyond doubt, but the evidence for it is sufficiently robust to merit sustained investigation. In this writer's view, ITC may be a hitherto unexplored domain of nonlocality, a form of nonlocal communication.
>
> To find a scientific basis for the ITC phenomenon, it must be connected with theories in physics. Transcommunication cannot be connected with classical physics, given that the latter is based on a paradigm that views phenomena not directly traceable to sensory

---

[3] Zammit and Zammit, A Lawyer Presents the Evidence for the Afterlife, p. 165.

experience as highly suspect, if not clearly illusory. But at the cutting edge of contemporary physics, many things and processes are acknowledged as elements of reality, even when they are intrinsically unobservable. Most pertinently, theories in particle as well as cosmological physics make reference to a field or dimension that subtends the world of the quantum, hitherto considered the lowest level of physical reality. This field or dimensions, variously termed "physical space-time," "nuether," "hyperspace," or "atemporal space" may be responsible for the phenomena of nonlocality in the microdomain of the quantum, as well as in the mesodomain of life and the macrodomain of the universe.[4]

Tesla, Marconi, Edison, and ITC

Professor at the Technical University (Fachhochschule) Bingen, Ernst Senkowski, Ph.D., of Mainz, Germany, defines ITC as "Technically supported contacts with 'beings, entities, Informational Structures' that are normally not accessible."[5] Dr. Senkowski states,

> Telecommunication devices (audio recorders, radio receivers, telephones, video recorders, TV receivers, computers) behave irregular[ly in] delivering voices, images, and texts of different, sometimes excellent quality coming from nowhere.
>
> The history of ITC: Tesla, Marconi, Edison suspected [it might be possible] to contact other worlds by electromagnetic means. Psychic mediums predicted the future development of suitable apparatus. Unexplained signals have been observed since WW I. Partially successful tests were realized in the 20s, 30s, 40s. First voices on tapes were documented independently during the 50s in Italy, USA, Sweden. TV-video images and computer texts appeared later. Official research is not known. Attempts to realize voices on tapes are carried out in at least 12 countries by single operators, small groups, and some bigger associations. Around 60 monographs have been published in

---

[4] Lazlo, "An unexplored domain of nonlocality: Toward a scientific explanation of Instrumental Transcommunication," http://bit.ly/1cTzBsY.

[5] Senkowski, "Instrumental Transcommunication ITC - in short," http://bit.ly/1cTzBsY.

seven languages. Only very few operators are scientifically interested. About 10 people have so far realized excellent results.

There are three open questions:

1. How are ITC events realized? Excellent results of these extraordinary psychophysical, mind-matter, man-machine, or consciousness-reality interactions apparently depend upon psychic faculties of the operators. A simple model using the parapsychological terms ESP [Extrasensory Perception] and PK [Psychokineses] describes ITC as a largely unconscious two-step process: a trans-entity from beyond our space-time system establishes a telepathic contact with a terrestrial operator who psychokinetically injects the trans-information into the electronic device. It cannot basically be excluded that our technical systems are directly manipulated from the "other side" using techniques unknown to us.

Understanding might result from: consciousness research, neurosciences, psychophysical theories, quantum theoretical models (Bohm, Jahn / Dunne-Pear / Princeton University); sub-quantum or vacuum physics; 12-dimenstional field theory (Burkhard Heim); theory of morphic fields (Sheldrake); system theories; OBE and NDE; thanatology; "far out" speculations [such as] scalar fields and tachyons.

2. What are the contents of ITC? It seems that some "No Body" wants to convince humanity about Life after Death in other states of consciousness and of the existence of nonhuman beings that could be involved to an uncontrollable extent. Generally [speaking], the contents confirm the fundamentals of mediumistic channeling in technical forms.

3. Who generates ITC? It seems unjustified to "explain" ITC phenomena as a mere result of the hypothetical "unconscious" of the operator or by super telepathy. Trans-entities appear autonomous and [appear] to be consciously alive. The selectivity of trans-information in cases of drop-in communicators cannot be explained and points to that direction. It is probable that the operator as an active transducer unconsciously adds a certain amount of his own associations to the primary message.

Important issues:

Demonstration of yet undeveloped faculties of the human mind.

Multidimensionality of humans.

Holistic interconnectedness and nonseparability of All-That-Is.

Mental awakening: opening windows to hitherto unknown realities or "parallel worlds," including "past" existences.

An ongoing and future metamorphosis or transformation of humanity.[6]

Relationship of ITC (Interlife) with Chronovision (Time Travel)

It is highly meaningful for the dimensional ecology of the Omniverse hypothesis that ITC's Electronic Voice Phenomena (EVP), used for exploring the Interlife in the spiritual dimensions, and chronovision, a time-travel technology for exploring timelines in universes of the Exopolitical dimensions, intersect in the work of two Italian Catholic priests, Fathers Pellegrino Ernetti and Augustino Gemelli.

Zammit writes,[7]

> [T]he Catholic Church has been actively positive and encouraging towards investigation of the Electronic Voice Phenomena:
>
>> Two of the earliest investigators were Italian Catholic priests, Father Ernetti and Father Gemelli, who came upon the phenomena by chance while they were recording Gregorian Chants in 1952. Father Gemelli heard his own father's voice on the tape, calling him by a childhood nickname, saying: "Zucchini, it is clear, don't you know it is I?"
>>
>> Deeply troubled by Catholic teaching in regard to contact with the dead, the two priests visited Pope Pius XII in Rome.
>>
>> The Pope reassured them: "Dear Father Gemelli, you really need not worry about this. The existence of this voice is strictly a

---

[6] Ibid.

[7] Zammit and Zammit, op. cit., p. 158.

scientific fact and has nothing to do with spiritism. The recorder is totally objective. It receives and records only sound waves from wherever they come. This experiment may perhaps become the cornerstone for a building for scientific studies which will strengthen people's faith in a hereafter."[8]

Pope Pius' cousin, the Rev. Professor Dr. Gebhard Frei, co-founder of the Jung Institute, was an internationally known parapsychologist who worked closely with Raudive, a pioneer in the research. He was also the President of the International Society for Catholic Parapsychologists. He himself is on record as stating, "All that I have read and heard forces me to believe that the voices come from transcendental, individual entities. Whether it suits me or not, I have no right to doubt the reality of the voices."[9]

Dr. Frei died on 27 October, 1967. In November 1967, at numerous taping sessions, a voice giving its name as Gebhard Frei came through. The voice was identified by Professor Peter Hohenwarter of the University of Vienna as positively belonging to Dr. Frei.[10]

Pope Paul VI was well aware of the work being done from 1959 onwards on the electronic voices by his good friend, Swedish film producer Friedrich Jürgenson, who had made a documentary film about him. The Pope made Jürgenson a Knight Commander of the Order of St. Gregory in 1969 for his work. Jürgenson wrote to Bander, "I have found a sympathetic ear for the Voice Phenomenon in the Vatican. I have won many wonderful friends among the leading figures in the Holy City. Today 'the bridge' stands firmly on its foundations."[11]

According to former US chrononaut Andrew D. Basiago, "The chronovisor was accidentally discovered when the Benedictine priests Pellegrino Ernetti and Augustino Gemelli were investigating the harmonic patterns in Gregorian chants at the Catholic University of Milan in the 1940's."[12] Basiago notes that Father

---

[8] Italian journal Astra, June 1990, quoted in Kubris and Macy, p. 102.
[9] Kubris and Macy, 1995.
[10] Ostrander and Schroeder, 1977.
[11] Ostrander and Schroeder, 1977. Zammit and Zammit, op. cit., pp. 158-159.
[12] Basiago, "The sources of time travel," http://bit.ly/1cPmciU.

Pellegrino Ernetti, discoverer of chronovision states, "From birth to death, every individual traces an arc of light and sound in the quantum hologram." By the "quantum hologram," Basiago and Father Pellegrino Ernetti are referring to dimensions in the universes of the Exopolitical dimension of the Omniverse.

Ernetti's Discovery of Instrumental Transcommunication (ITC) with the Interlife

The story of Ernetti's original discovery of the Instrumental Transcommunication (ITC) capabilities of his machine through interactive communication with his deceased father, and a translation of a 1972 Italian newspaper article, "Invention of a machine for photographing the past," establish at least two applications of the ITC/chronovisor as developed by Ernetti and Gemelli. The chronovisor of Ernetti and Gemelli was used to achieve:

- Instrumental Transcommunication (ITC) resulting in interactive communication with the intelligences of specific deceased humans located in the Interlife in the spiritual dimensions of the Omniverse; and

- Chronovision Time Travel: Location and video and audio recording of events along a past timeline in the time-space hologram of the Multiverse.

Thus, it appears that by using a common instrument, the chronovisor as invented by Ernetti and Gemelli, the existence of time travel along timelines in the time-space hologram (quantum hologram) in the Multiverse can be established through their reported chronovision recording of famous personages in history.

Ernetti himself states how the chronovisor can tune into individual voices and images along a timeline in the time-space hologram of the Exopolitical dimension, with the use of suitable equipment, [which] our team was the first in the world to build. The equipment consists of a series of antennas to allow tuning of the individual voices and images. We know that every human being, from when he was born to when he dies, leaves behind himself a double wake: a visual and a sound, a sort of identity card for each different person. And based on this different identity card one can reconstruct the single person in all his deeds and his sayings, and for this reason we are now able to hear and see again the greatest people in history.[13]

Ernetti's invention is also capable of Instrumental Transcommunication (ITC), which is interactive communication with the intelligence of an individual soul that was formerly incarnated in the time-space hologram of the Exopolitical dimensions and, at the time of ITC, may be based in the Interlife of the spiritual dimensions. If the individual has not yet "crossed over" or teleported through the interdimensional portal separating the Exopolitical and spiritual dimensions, the Instrumental Transcommunication (ITC) will be with the soul of an individual located in an Exopolitical dimension awaiting transfer into the Interlife.

Ernetti and Gemelli's most cogent example of Instrumental Transcommunication (ITC) occurs in their report that they achieved interactive ITC with the intelligence of Ernetti's deceased father, located in the Interlife dimension of the spiritual dimensions, calling Ernetti "by a childhood nickname, saying, *'Zucchini, it is clear, don't you know it is I?'"*

Ernetti and Gemelli's experiments provide empirical evidence supporting the following:

1 The dimensional ecology hypothesis,
2 That we humans exist in a time-space hologram
3 As incarnating avatars in the Exopolitical dimensions

---

[13] Maddaloni, "Invention of a machine for photographing the past," http://on.fb.me/HykL0u.

4   For souls that are based in the Interlife in the spiritual dimensions

5   Of an Omniverse that consists of both Exopolitical and spiritual dimensions.

REINCARNATION STUDIES

Prima facie empirical evidence for the dimensional ecology hypothesis and for existence of the intelligent civilizations of souls based in the Interlife of the spiritual dimensions has been developed through science-based studies suggestive of reincarnation. Reincarnation is a fundamental process in the design of the dimensional ecology of the Omniverse. By means of it, a soul based in the Interlife dimensions of the spiritual dimensions can transition into a parallel dimension or universe of the Multiverse and can undertake an incarnated life, using the support system of the physical body of some creature or entity as an avatar, usually in an intelligent civilization. Some spiritual traditions speak of the transmigration of human souls into the animal, vegetable, or mineral kingdoms as well. Our focus here is on the study of reincarnation of souls in the dimensional ecology of intelligent civilizations.

The two principal methods of scientific scrutiny of reincarnation in the Exopolitical dimensions are hypnotic regression to capture memories of prior (and future) lives, and study of spontaneous memories of past lives among children. We will focus on hypnotic regression in reconstructing the nature of the afterlife or Interlife dimension.

SPONTANEOUS MEMORIES OF PAST LIVES IN THE DIMENSIONAL ECOLOGY

Dr. Ian Stevenson (1918-2007), formerly of the University of Virginia Medical School, studied more than 3,000 cases suggestive of reincarnation, in children who had spontaneous memories of apparent prior lives in the dimensional ecology. These cases provide proof of the continuation of consciousness after physical death, because each case provides empirically verified evidence of the continuation of an underlying consciousness (soul) in a new physical body.[14]

There are common patterns in all these cases, in that the children who have spontaneous memories of prior lives began relating facts and stories about their prior lives between the ages of 2 and 4 years old. By the age of 8 or 9 years old, the children by and large appeared to have forgotten about the prior lives. Other patterns in these cases are the frequent mention by the children of having died a violent death in the prior life, and a clear memory of the mode of death.

Memories of a past life

A summary of Dr. Ian Stevenson's cases demonstrates patterns of *prima facie* evidence supporting the dimensional ecology hypothesis. In most of Dr. Stevenson's cases, the same underlying soul is demonstrated as carrying memories from a prior incarnation as an Earthling human in the time-space hologram of the Exopolitical dimensions. One summary states,

> The most frequently occurring event or common denominator relating to rebirth is probably that of a child remembering a past life. Children usually begin to talk about their memories between the ages of two and four. Such infantile memories gradually dwindle when the child is between four and seven years old. There are, of course, always some exceptions, such as a child continuing to remember its previous life but not speaking about it for various reasons.
>
> Most of the children talk about their previous identity with great intensity and feeling. Often they cannot decide for themselves which

---

[14] For an overview of Stevenson's work, see University of Virginia Medical School, The Division of Perceptual Studies, http://bit.ly/17cJkMo

world is real and which one is not. They often experience a kind of double existence, where at times one life is more prominent, and at times the other life takes over. This is why they usually speak of their past life in the present tense, saying things like, 'I have a husband and two children who live in Jaipur.' Almost all of them are able to tell us about the events leading up to their death.

Such children tend to consider their previous parents to be their real parents rather than their present ones, and usually express a wish to return to them. When the previous family has been found and details about the person in that past life have come to light, then the origin of the fifth common denominator—the conspicuous or unusual behavior of the child—is becoming obvious.

For instance, if the child is born in India to a very low-class family and was a member of a higher caste in its previous life, it may feel uncomfortable in its new family. The child may ask to be served or waited on hand and foot and may refuse to wear cheap clothes. Stevenson gives us several examples of these unusual behavior patterns.[15]

Birthmarks and wounds suffered in a prior life

Another empirical regularity Stevenson demonstrates, one that is *prima facie* evidence, is that a specific soul may manifest wounds suffered in a prior life as birthmarks in a present life in the Earth's time-space hologram in the Exopolitical dimensions.

Stevenson writes,

> Almost nothing is known about why pigmented birthmarks (moles or nevi) occur in particular locations of the skin. The causes of most birth defects are also unknown. About 35 percent of children who claim to remember previous lives have birthmarks and/or birth defects that they (or adult informants) attribute to wounds on a person whose life the child remembers. The cases of 210 such children have been investigated. The birthmarks were usually areas of hairless, puckered skin; some were areas of little or no pigmentation (hypopigmented

---

[15] Reincarnation Research, http://bit.ly/HKqjF9

macules); others were areas of increased pigmentation (hyperpigmented nevi). The birth defects were nearly always of rare types. In cases in which a deceased person was identified, the details of whose life unmistakably matched the child's statements, a close correspondence was nearly always found between the birthmarks and/or birth defects on the child and the wounds on the deceased person. In 43 of 49 cases in which a medical document (usually a postmortem report) was obtained, it confirmed the correspondence between wounds and birthmarks (or birth defects). There is little evidence that parents and other informants imposed a false identity on the child in order to explain the child's birthmark or birth defect. Some paranormal process seems required to account for at least some of the details of these cases, including the birthmarks and birth defects.

Because most (but not all) of these cases develop among persons who believe in reincarnation, we should expect that the informants for the cases would interpret them as examples according with their belief; and they usually do. It is necessary, however, for scientists to think of alternative explanations.

The most obvious explanation of these cases attributes the birthmark or birth defect on the child to chance, and the reports of the child's statements and unusual behavior then become a parental fiction intended to account for the birthmark (or birth defect) in terms of the culturally accepted belief in reincarnation. There are, however, important objections to this explanation.

First, the parents (and other adults concerned in a case) have no need to invent and narrate details of a previous life in order to explain their child's lesion. Believing in reincarnation, as most of them do, they are nearly always content to attribute the lesion to some event of a previous life without searching for a particular life with matching details.

Second, the lives of the deceased persons figuring in the cases were of uneven quality both as to social status and commendable conduct. A few of them provided models of heroism or some other enviable quality; but many of them lived in poverty or were otherwise unexemplary. Few parents would impose identification with such persons on their children.

Third, although in most cases the two families concerned were acquainted (or even related), I am confident that in at least 13 cases (among 210 carefully examined with regard to this matter) the two families concerned had never even heard about each other before the case developed. The subject's family in these cases can have had no information with which to build up an imaginary previous life that, it later turned out, closely matched a real one. In another 12 cases the child's parents had heard about the death of the person concerned, but had no knowledge of the wounds on that person ... .

Fourth, I think I have shown that chance is an improbable interpretation for the correspondences in location between two or more birthmarks on the subject of a case and wounds on a deceased person.

Persons who reject the explanation of chance combined with a secondarily confected history may consider other interpretations that include paranormal processes, but fall short of proposing a life after death. One of these supposes that the birthmark or birth defect occurs by chance and the subject then by telepathy learns about a deceased person who had a similar lesion and develops identification with that person. The children subjects of these cases, however, never show paranormal powers of the magnitude required to explain the apparent memories in contexts outside of their seeming memories.

Another explanation, which would leave less to chance in the production of the child's lesion, attributes it to a maternal impression on the part of the child's mother. According to this idea, a pregnant woman, having knowledge of the deceased person's wounds, might influence a gestating embryo and fetus so that its form corresponded to the wounds on the deceased person. The idea of maternal impressions, popular in preceding centuries and up to the first decades of this one, has fallen into disrepute. Until my own recent article (Stevenson, 1992) there had been no review of series of cases since 1890 ... Nevertheless, some of the published cases—old and new—show a remarkable correspondence between an unusual stimulus in the mind of a pregnant woman and an unusual birthmark or birth defect in her later-born child. Also, in an analysis of 113 published cases I found that the stimulus occurred to the mother in the first trimester in 80 cases (Stevenson, 1992). The first trimester is well known to be the one of greatest sensitivity of the embryo/fetus to recognized teratogens, such as thalidomide ... Applied to the present cases, however, the theory of maternal impression has obstacles as great as the normal explanation appears to have. First, in the 25 cases mentioned above, the subject's mother, although she may have heard of the death of the concerned deceased person, had no knowledge of that person's wounds. Second, this interpretation supposes that the mother not only modified the body of her unborn child with her thoughts, but also after the child's birth influenced it to make statements and show behavior that it otherwise would not have done. No motive for such conduct can be discerned in most of the mothers (or fathers) of these subjects.[16]

NEAR DEATH EXPERIENCES (NDES)

Other methodologies exist for developing replicable empirical evidence supporting the dimensional ecology hypothesis of the Omniverse. For example, David Fontana lists a total of seventeen formal methodologies for proving the existence of a dimensional ecology including the Interlife dimensions.[17] Notable among these methodologies is the study of Near Death Experiences (NDEs), which provide *prima facie* eyewitness evidence of the dimensional ecology, of the transition of the human soul through the dimensional ecology, and of the spiritual dimension itself.[18]

Evidence that Near-Death Experiences (NDEs) are real

Victor Zammit has summarized fourteen points to show why NDEs are real and not hallucinations produced by lack of brain oxygen or drugs. He writes,

1 NDE survivors have clear and structured memories of what happened to them. Patients who did not have a NDE during similar treatment were confused or could not remember anything.

---

[16] Reincarnation Research, http://bit.ly/HKqjF9.

[17] See David Fontana, Is There An afterlife? A Comprehensive Overview of the Evidence, and Life Beyond Life; What Should We Expect.

[18] On this topic see especially Kenneth Ring, Lessons from the Light: What We Can Learn from Near Death Experiences, and Raymond A. Moody, Jr., Life After Life.

2 Whereas hallucinations are all different, near-death experiences are very similar in different cultures and throughout history Near-death experiences have been reported in all cultures, and from as far back as 1760 BC ... .

3 People see and hear things while they are unconscious that would be impossible for normal sensing. A huge percentage of near-death experiencers are able to describe exactly what happened to them while they were unconscious.

4 People come back from a NDE with accurate facts they did not know before.

5 People report meeting with relatives they did not know were dead. In all cases they are correct.

6 Some people come back with knowledge of the future.

7 Some people come back with advanced knowledge consistent with quantum physics. Almost all survivors say that they entered a dimension where there was no time and many were able to go back and forward through time.

8 Some people are cured of fatal illnesses during a NDE or have miraculous recoveries from serious injuries.

9 The blind can see during a NDE.

10 Some people have a group near-death experience. A group of fire fighters claimed that when they were overcome with smoke, they all went out of their bodies. They communicated with each other and could all see the lifeless bodies below them.

11 Some people have near-death-like experiences when there is nothing physically wrong with them. Researchers have found that deep meditation, deathbed visions, relaxation, psychic vision, astral projection, trance, mirror-gazing, and eye movements can trigger elements of the NDE.

12 Some people have a near-death experience when they are completely brain dead. Hallucinations can only occur when people have a functioning brain which shows an active EEG reading.

13 Many people experience a "life review" during which they see their lives from the perspective of other people.

14 The after-effects of a NDE are unique and long lasting.

There have been many attempts to explain away the near-death experience. Some claim they are caused by oxygen deprivation. Others claim it is a natural effect of the dying brain or some accident of brain chemistry. Most of these theories are based on observations of a small number of cases. They may produce elements of the near-death experience but not the whole experience. And, most important, they do not have the same impact or after-effects.[19]

Cross-correspondence Experiments Through Physical Mediums

Another notable methodology is controlled experiments involving physical mediumship. One experiment using physical mediums to develop *prima facie* evidence of the dimensional ecology and the human soul's survival in the Interlife in the spiritual dimensions was that of cross-correspondences, designed by Cambridge classics scholar Frederick W.H. Myers, who after he died transmitted a series of signed scripts though more than a dozen mediums in different parts of the world.

Zammit writes,

> Later, there were scripts signed by his fellow leaders of the Society for Psychical Research, Professor Henry Sidgwick and Edmund Gurney,

---

[19] Zammit and Zammit, op. cit., pp. 22-30.

as they too had died. The scripts were all about unusual classical subjects and did not make sense on their own. But the mediums were told to contact a central address and send the scripts there. When the scripts were put together they fitted like pieces of a jigsaw puzzle. In all, more than 3,000 scripts were transmitted over 30 years. Some of them were more than 40 typed pages long. Together they fill 24 volumes and 12,000 pages. The investigation went on so long that some of the investigators, such as Professor Verrall, died during the course of it and began communicating themselves.[20]

Biocentrism and the Dimensional Ecology

Even scientist Robert Lanza's Biocentrism holds that the following is supportive of a dimensional ecology in what it perceives as the universes of the multiverse.

> There is no separate physical universe outside of life and consciousness. Nothing is real that is not perceived. There was never a time when an external, dumb, physical universe existed, or that life sprang randomly from it at a later date. Space and time exist only as constructs of the mind, as tools of perception. Experiments in which the observer influences the outcome are easily explainable by the interrelatedness of consciousness and the physical universe. Neither nature nor mind is unreal; both are correlative. No position is taken regarding God.[21]

One analysis of biocentrism concludes

> The theory [Biocentrism] implies that death simply does not exist. It is an illusion which arises in the minds of people. It exists because people identify themselves with their body. They believe that the body is going to perish, sooner or later, thinking their consciousness will disappear too. In fact, consciousness exists outside of constraints of time and space. It is able to be anywhere: in the human body and outside of it. That fits well with the basic postulates of quantum mechanics science, according to which a certain particle can be present anywhere and an event can happen according to several, sometimes countless, ways.

---

[20] Zammit and Zammit, op. cit., pp. 110-111.
[21] Berman and Lanza, Biocentrism, p. 159.

Lanza believes that multiple universes can exist simultaneously. These universes contain multiple ways for possible scenarios to occur. In one universe, the body can be dead. And in another it continues to exist, absorbing consciousness which migrated into this universe.

This means that a dead person while traveling through the same tunnel ends up not in hell or in heaven, but in a similar world he or she once inhabited, but this time alive. And so on, infinitely....

So, there is an abundance of places or other universes where our soul could migrate after death, according to the theory of neo-biocentrism. But does the soul exist?

Professor Stuart Hameroff from the University of Arizona has no doubts about the existence of eternal soul. As recently as last year, he announced that he has found evidence that consciousness does not perish after death.

According to Hameroff, the human brain is the perfect quantum computer and the soul or consciousness is simply information stored at the quantum level. It can be transferred, following the death of the body; quantum information represented by consciousness merges with our universe and exists there indefinitely. The biocentrism expert Lanza proves that the soul migrates to another universe. That is the main difference from his other colleagues.

Sir Roger Penrose, a famous British physicist and expert in mathematics from Oxford, supports this theory, and he has also found traces of contact with other universes. Together, the scientists are developing quantum theory to explain the phenomenon of consciousness. They believe that they found carriers of consciousness, the elements that accumulate information during life and [that], after death of the body ... "drain" consciousness somewhere else. These elements are located inside protein-based microtubules (neuronal microtubules), which previously have been attributed a simple role of reinforcement and transport channeling inside a living cell. Based on their structure, microtubules are best suited to function as carriers of quantum properties inside the brain ... because they are able to retain quantum states for a long time [and can thus] function as elements of a quantum computer.[22]

---

[22] Learning-Mind.com, "Quantum theory proves that consciousness moves to another universe after death," http://bit.ly/1f6wQFk.

Dimensional Ecology Between the Exopolitical Dimensions and the Interlife Dimensions

We have reviewed *prima facie* evidence supporting the existence of Interlife dimensions, where the intelligent civilizations of souls is based between experiential incarnations in diverse forms in the universes of the Exopolitical dimensions (multiverse) of the Omniverse.

This *prima facie* evidence also describes instances of a dimensional ecology that exists between the intelligent civilizations in the Exopolitical dimensions and intelligent civilizations of souls based in the Interlife. As we have seen, the evidence for dimensional communication between the Exopolitical dimensions and the Interlife includes:

- Instrumental Transcommunication (ITC) and Electronic Voice Phenomena
- Reincarnation studies
- Near-death experiences (NDEs)
- Cross-correspondence experiments through physical mediums

CHAPTER 7
THE INTELLIGENT CIVILIZATIONS OF SOULS

The life of souls in the dimensional ecology has long been recorded in human history. Anthony Peake tells of Plato's descriptions of the soul in the dimensional ecology of the Omniverse. Peake writes, "Plato is clear that the soul survives after death. He describes, through the experiences of [the soul of] Er, that some form of judgment takes place. This is made by celestial beings who, although it is not stated explicitly, seem to review the life of the soul in question. A decision is then made as to whether the soul is reborn again ('to the left and downward') or is allowed to move on to a higher plane of existence. This image we will see described again and again."[1]

Replicable data from hypnotic regression of soul memories of the Interlife now provide *prima facie* empirical evidence supporting the dimensional ecology of the Omniverse hypothesis by giving us detailed information about the interactions of the intelligent civilizations of souls, of spiritual beings, and of Source (God) with the universes and intelligent civilizations of the

---

[1] Peake, Is there Life after Death? Kindle locations 5039-5040.

Exopolitical dimensions (multiverse).[2]

During 1960 to 1980, Michael Newton developed and applied standard protocols in using hypnotic regression to access replicable soul memories of the dimensional ecology in the Interlife dimensions. His data eventually grew to more than 7,000 replicable cases, in a body of work that was initially published in 1995, more than thirty years after it was commenced.[3]

NATURE AND CREATION OF SOULS AS HOLOGRAPHIC FRAGMENTS OF GOD

Given his observations of more than 7,000 replicable cases, Newton defines the soul as "intelligent light energy." He writes,

> We cannot define the soul in a physical way because to do so would establish limits on something that seems to have none. I see the soul as intelligent light energy. This energy appears to function as vibrational waves similar to electromagnetic force but without the limitations of charged particles of matter. Soul energy does not appear to be uniform. Like a fingerprint, each soul has a unique identity in its formation, composition, and vibrational distribution. I am able to discern soul properties of development by color tones, yet none of this defines what the soul is as an entity.[4]

Newton's replicable data expands on the creation of souls as holographic "eggs of Light" from God, who manifests as a "Sea of Light." In this regression, Newton is speaking with a soul that works at a soul-creation facility in the spiritual dimensions.

Newton: As an Incubator Mother, when do you first see the new souls?

Soul: We are in the delivery suite, which is a part of the nursery, at one end of the emporium. The newly arrived ones are conveyed as small masses of white energy encased in a gold sac. They move slowly in a majestic, orchestrated line of progression toward us.

---

[2] See Newton's books in the Sources and Resources list.
[3] Miejan, "Journey of Souls with Michael Newton," http://bit.ly/1bpsY2m.
[4] Newton, Destiny of Souls, Kindle locations 1566-1567.

Dr. Newton: From where?

Soul: At our end of the emporium under an archway the entire wall is filled with a molten mass of high-intensity energy and … vitality. It feels as if it's energized by an amazing love force rather than a discernible heat source. The mass pulsates and undulates in a beautiful flowing motion. Its color is like that on the inside of your eyelids if you were to look through closed eyes at the sun on a bright day.

Dr. Newton: And from out of this mass you see souls emerge?

Soul: From the mass a swelling begins, never exactly from the same site twice. The swelling increases and pushes outward, becoming a formless bulge. The separation is a wondrous moment. A new soul is born. It's totally alive with an energy and distinctness of its own.[5]

Newton added the following note from another interview with a high-level soul.

> Another one of my [high] level V [souls] made this statement about incubation. "I see an egg-shaped mass with energy flowing out and back in. When it expands, new soul energy fragments are spawned. When the bulge contracts, I think it pulls back those souls which were not successfully spawned. For some reason these fragments could not make it on to the next step of individuality.[6]

SOULS IN THE INTERLIFE DIMENSIONS

The replicable findings of Dr. Newton and his associates on the intelligent civilizations of souls in the Interlife dimensions include:

1. The nature and origin of souls as holographic eggs of Light from the Sea of Light that is the Creator Source (God);

2. The origin and function of the universes of the Exopolitical Dimensions as a virtual experiential development environment for souls;

---

[5] Newton, op. cit., p. 127.
[6] Newton, op. cit., Kindle locations 2261-2269.

3   The development and education of souls through groups in the Interlife;

4   The process of soul development through incarnation into a physical body or avatar in a universe of the Exopolitical dimensions as an intelligent species—human, mammalian, reptilian, avian, etc.

5   The specific steps in the incarnation process include:
    - Key roles of Guides, trained souls assigned to each incarnating soul;
    - Selecting a future physical life, including technological "sampling of lives";
    - Rehearsal and recognition classes for chosen future lives with incarnating soul mates;
    - Teleportation (in a normal incarnation and birth) through an interdimensional portal from the Interlife dimensions into the Exopolitical dimensions to coordinates next to the womb of the prospective mother, enmeshing into the fetus of the soul's new life about three months prior to birth (final trimester);
    - In unusual cases, a single soul may lead simultaneous, different physical lives in different locations of the Exopolitical dimensions;
    - In normal cases, at physical death, the soul teleports back through an interdimensional portal into the Interlife, and (unless an advanced soul) is met by guides and family, friends, etc.
    - A life-review period follows with guides and a Council of Elders, including holographic review of life episodes for karmic lessons, etc.;

- The soul enters a transition period for further education, new soul career as guide, healer, new life planning, etc.;
- Placement of the soul in this new area of activity, and, where necessary,
- Start of a new life-selection process.

DIMENSIONAL ECOLOGY: REENTRY FROM THE EXOPOLITICAL DIMENSIONS INTO THE INTERLIFE DIMENSIONS

Replicable data from hypnotic regression of soul memories of the Interlife set out standard treatment for souls that are in the dimensional ecology, re-entering the interdimensional portal to the spiritual Interlife dimensions. The soul guides use three techniques at the interdimensional portal:

1. Envelopment. Entering souls are enveloped in a large circular mass of energy by their guide to ease the transition from incarnation in the Exopolitical dimension.

2. Focus effect. The guide can apply energy to border points, as a sort of spiritual acupuncture, along meridians of the soul's etheric body;

3. Emergency treatment at the interdimensional portal to the spiritual dimensions. If an arriving soul has experienced the violent death of the physical body it was incarnated in, the soul may arrive with its energy in a deteriorated state. The guides will then give energetic and spiritual mediation exercises to the soul before it moves forward into the Interlife and the spiritual dimensions.[7]

---

[7] Newton, op. cit., p. 86.

Soul Groups in the Interlife Dimensions

Replicable data from hypnotic regression of soul memories of the Interlife indicate that souls are assigned to soul groups with which they tend to remain, in a context of total freedom. In reincarnation, soul groups may tend to navigate the dimensional ecology together, incarnated as groups in varying social roles on planets in the universes of the Exopolitical dimensions.

1. New souls are assigned to a new soul support group based on their level of understanding.

2. Once a new soul support group is formed, new soul members are not added in the future.

3. There is a process for systematic selection of homogeneous groups of souls. Similarities in ego, cognitive consciousness, expression, and desire are defining considerations;

4. Regardless of size, soul groups do not mix their energy with that of other soul groups. Souls can communicate between themselves.

5. Learning curves can differ for individual souls. Some souls advance more rapidly than others. At the intermediate level, should they show special talents (healing, creativity, teaching, etc.), they are authorized to participate in special training while still in their soul groups.

6. At the intermediate level, souls may join together in large independent study groups.

7. At the more advanced level, souls can undertake independent activities outside the group and come to act as guides.

8. classification model for soul development has found the following mix of soul development levels and auras:

Beginner's souls, 42 percent (white aura);
   Intermediate souls, 57 percent (yellow-gold aura);
   Advanced, elevated souls, 1 percent (light blue-purple aura).[8]

Dimensional Ecology: Life of a Soul Incarnated as an Extraterrestrial

Empirical *prima facie* evidence exists, supporting the dimensional ecology hypothesis, that souls from the Interlife in the spiritual dimensions incarnate (by teleportation) as physical lives in the universes of the Exopolitical dimensions. Newton states, "I have an extensive file on souls that have incarnated in other worlds and souls that have traveled in a variety of unknown worlds for study and recreation between their lives on Earth."[9]

Dimensional Ecology: Time Travel, Teleportation, and Interdimensional Portals Between Exopolitical and Interlife Dimensions

The replicable data reported by Newton and his associates using standard protocols in hypnotic regression, about soul memories of teleportation through the dimensional ecology between the Exopolitical and Interlife dimensions of the Omniverse, are congruent with a report by Andrew D. Basiago. Basiago reported to me on his seeing and hearing recently deceased souls while time-travel teleporting as a US chrononaut through a vortal tunnel in the time-space hologram of our universe that was created by secret time-travel technology of the US government. These souls were in dimensions or "astral planes" adjacent to the vortal tunnel that Basiago was teleporting through. Apparently these souls were still in transition in the time-space hologram of

---

[8] Newton, Journey of Souls, p. 103.
[9] Newton, Destiny of Souls, p. 340. .

our universe, on their way to the interdimensional portal to the Interlife dimensions.

> One of the Newton's replicable findings describes a dimensional ecology like that encountered by Basiago in the dimensional ecology immediately surrounding Earth. Newton writes,

> Once through the tunnel, our souls have passed the initial gateway of their journey into the spirit world. Most now fully realize they are not really dead, but have simply left the encumbrance of an Earth body which has died. With this awareness comes acceptance in varying degrees depending upon the soul. Some subjects look at these surroundings with continued amazement while others are more matter-of-fact in reporting to me what they see. Much depends upon their respective maturity and recent life experiences. The most common type of reaction I hear is a relieved sigh followed by something on the order of, "Oh, wonderful, I'm home in this beautiful place again." There are those highly developed souls who move so fast out of their bodies that much of what I am describing here is a blur as they home into their spiritual destinations. These are the pros and, in my opinion, they are a distinct minority on Earth. The average soul does not move that rapidly and some are very hesitant. If we exclude the rare cases of highly disturbed spirits who fight to stay connected with their dead bodies, I find it is the younger souls with fewer pasts.

> Most of my subjects report that as they emerge from the mouth of the tunnel, things are still unclear for awhile. I think this is due to the density of the nearest astral plane surrounding Earth, called the *kamaloka* by Theosophists.[10]

DIMENSIONAL ECOLOGY: LIFE CREATION IN THE UNIVERSES OF THE EXOPOLITICAL DIMENSIONS

Replicable data developed by Newton and his associates about the dimensional ecology between the intelligent civilizations of souls in the Interlife dimensions and in the universes in the Multiverse illustrate the roles that the intelligent civilizations of souls have

---

[10] Newton, Journey of Souls, Kindle locations 261-265, 287-294, 294-296.

played in creating life in the multiverse. These roles are consistent with the replicable finding that "God" is the totality of the spiritual dimension of the Omniverse, namely, the intelligent civilizations of souls, of spiritual beings, and God itself.

In this transcript, Newton illustrates how a soul can participate in creating life:

Dr. Newton: Let me try and sum this up, and please tell me if I am on the wrong track. A soul who becomes proficient with actually creating life must be able to split cells and give DNA instructions, and you do this by sending particles of energy into protoplasm?

Soul: We must learn to do this, yes, coordinating it with a sun's energy.

Dr. N: Why?

S: Because each sun has different energy effects on the worlds around them.

Dr. N: Then why would you interfere with what a sun would naturally do with its own energy on a planet?

S: It is not interference. We examine new structures ... mutations ... to watch and see what is workable. We arrange substances for their most effective use with different suns.

Dr. N: When a species of life evolves on a planet, are the environmental conditions for selection and adaptation natural, or are intelligent soul-minds tinkering with what happens?

S: (*evasively*) Usually a planet hospitable to life has souls watching and whatever we do is natural.

Dr. N: How can souls watch and influence biological properties of growth evolving over millions of years on a primordial world?

S: Time is not in Earth years for us. We use it to suit our experiments.

Dr. N: Do you personally create suns in our universe?

S: A full-scale sun? Oh no, that's way over my head ... and requires the powers of many. I generate only on a small scale.

Dr. N: What can you generate?

S: Ah ... small bundles of highly concentrated matter ... heated.

Dr. N: But what does your work look like when you are finished?

S: Small solar systems.

Dr. N: Are your miniature suns and planets the size of rocks, buildings, the moon—what are we talking about here?

S: (*laughs*) My suns are the size of basketballs and the planets ... marbles ... that's the best I can do.

Dr. N: Why do you do this on a small scale?

S: For practice, so I can make larger suns. After enough compression the atoms explode and condense, but I can't do anything really big alone.

Dr. N: What do you mean?

S: We must learn to work together to combine our energy for the best results.

Dr. N: Well, who does the full-sized thermonuclear explosions which create physical universes and space itself?

S: The Source ... the concentrated energy of the Old Ones.

Dr. N: Oh, so the Source has help?[11]

NATURE OF SOURCE (GOD) AND DIMENSIONAL ECOLOGY BETWEEN EXOPOLITICAL AND SPIRITUAL DIMENSIONS

Newton has developed data on the nature of Source (God) and on the dimensional ecology between the universes of the Exopolitical dimensions and the spiritual dimensions (souls, spiritual beings, and God itself). The data support the dimensional ecology of the Omniverse hypothesis by demonstrating how the dimensional ecology operates between a spiritual dimension that (1) consists of the intelligent civilizations of souls, of spiritual beings, and of God itself; and (2) functionally is the Source of the created universes of

---

[11] Newton, op. cit., Kindle locations 2325-2326.

time, space, energy, and matter in the Exopolitical dimensions (multiverse).

> *Dr. Newton:* Tell me, does the Source dwell in some special central space in the spirit world [spiritual dimensions]?
>
> *Soul:* The Source is the spirit world [spiritual dimensions].[12]
>
> *Dr. N:* If the Source represents all the spirit world, how does this mental place differ from physical universes with stars, planets, and living things?
>
> *S:* Universes are created—to live and die—for the use of the Source. The place of spirits ... is the Source.
>
> *Dr. N:* We seem to live in a universe which is expanding and may contract again and eventually die. Since we live in a space with time limitations, how can the spirit world itself be timeless?
>
> *S:* Because here we live in non-space which is timeless ... except in certain zones.
>
> *Dr. N:* Please explain what these zones are.
>
> *S:* They are ... interconnecting doors ... openings for us to pass through into a physical universe of time.
>
> *Dr. N:* You speak of universes in the plural. Are these other physical universes besides the one which contains Earth?
>
> *S:* (*vaguely*) There are ... differing realities to suit the Source.
>
> *Dr. N:* Are you saying souls can enter various rooms of different physical realities from spiritual doorways?
>
> *S:* (*nods*) Yes, they can—and do.[13]

---

[12] Ibid., Kindle Locations 2325-2326.

[13] Ibid., Kindle locations 2330-2334, 2339-2341.

Dimensional Ecology: Soul Creates Galactic Matter, Stars, and Planets in the Exopolitical Dimensions

In regression interviews with an advanced soul of the "Structural Soul" category, Newton was able to establish the role of advanced souls in creating the intelligent civilizations of souls in the dimensional ecology. These advanced "Structural Souls" create galactic matter, stars, and planets in the universes of the Exopolitical dimensions. This replicable data further supports the dimensional ecology of the Omniverse hypothesis.

Newton writes,

> My subjects who are Structural Souls say these designs relate to the formation of "geometric shapes that float as elastic patterns," which contribute to the building blocks of a living universe.... The Masters of Design have enormous influence on creation. I'm told they are capable of bridging universes that seem not to have a beginning or end, exacting their purposes among countless environmental settings. Carried to its logical conclusion, this would mean these masters—or grandmasters—would be capable of creating the spinning gas clouds of galactic matter which started the process of stars, planets.[14]

Dimensional ecology: Interdimensional travel to create or adjust planets in a universe

Replicable data from hypnotic regression from soul memories of the Interlife that support the dimensional ecology hypothesis also reveal that advanced souls of the "Explorer" category travel interdimensionally from the Interlife dimensions to the universes of the Exopolitical dimensions (multiverse) while they are creating or adjusting planets within a universe in the Exopolitical dimensions.

---

[14] Newton, Destiny of Souls, Kindle locations 5615-5620.

Newton writes,

> Souls who travel interdimensionally [into the universes of the Exopolitical dimensions] explain that their movements appear to be in and out of curved spheres connected by zones that are opened and closed by converging vibrational attunement. Explorer [soul] trainees have to learn this skill. From the accounts I have heard, the interdimensional [soul] travelers must also learn about the surface boundaries of zones as hikers locating trailheads between mountain ranges. Souls speak of points, lines, and surfaces in multi-space which indicate larger structural solids, at least for the physical universes. I would think dimensions having geometric designs need hyperspace to hold them. Yet Explorer souls travel so fast in some sort of hyperspace it seems to me the essence of speed, time, and direction of travel is hardly definitive. Training to be an Explorer [soul] must indeed be formidable, as indicated by a quote from this client [whose soul] travels through five dimensions between her lives:

> These dimensions are so enmeshed with one another that I have no sense of boundaries except for two elements, sound and color. I must learn to attune my energy to the vibrational frequency of each dimension, and some are so complex I cannot yet go to them. With color, the purples, blues, yellows, reds and whites are manifestations of light and density for those energy particles in the dimensions where I travel.[15]

DIMENSIONAL ECOLOGY: INCARNATIONS TO ALTER TIME IN THE TIME-SPACE HOLOGRAM

Data from hypnotic regression from soul memories of the Interlife supporting the dimensional ecology hypothesis also reveal incarnations on simulated planets in the Exopolitical dimensions for purposes of soul development of specific skills, such as mastery of the time dimension in the time-space holograms in the universes of the multiverse. Another of Newton's transcripts reads:

---

[15] Ibid., p. 354.

Dr. Newton: And you don't physically live on this simulated world which appears as Earth—you only use it?

Soul: Yes, that's right, for training purposes.

Dr. N: Why do you call this third sphere the World of Altered Time?

S: Because we can change time sequences to study specific events.

Dr. N: What is the basic purpose of doing this?

S: To improve my decisions for life.[16]

DIMENSIONAL ECOLOGY: SOUL INTERVENTIONS IN DREAMS OF INDIVIDUALS IN THE MULTIVERSE

Replicable regression/interviews with advanced souls of the "Dreamweaver" category by Newton indicate their role in influencing interdimensional consciousness in the dimensional ecology, specifically that of dreams of persons in our universe. Newton writes,

The Dreamweaver souls I have come in contact with all engage in dream implanting, with two prominent differences.

1. Dream Alteration. Here a skillful discarnate [soul] enters the mind of a sleeper and partially alters an existing dream already in progress. This technique I would call one of interlineation, where spirits place themselves as actors between the lines of an unfolding play so the dreamer is not aware of script tampering with the sequences .... As difficult as this approach seems, it is evident to me the second procedure is more complex.

2. Dream Origination. In these cases the soul must create and fully implant a new dream from scratch and weave the tapestry of these images into a meaningful presentation to suit their purpose. Creating or altering scenes in the mind of a dreamer is intended to convey a message. I see as this an act

---

[16] Newton, Journey of Souls, p. 160.

of service and love. If the dream implantation is not performed skillfully to make the dream meaningful, the sleeper moves on and wakes up in the morning remembering only disjointed fragments or nothing at all about the dream.[17]

The "Council of Elders" and the role of Grey extraterrestrials in a soul's reincarnation cycle

Researcher David Wilcock speculates that the "Council of Elders" that Dr. Michael Newton's replicable research found as a consistent feature of the Interlife experience may in fact be composed of extraterrestrials, perhaps of the Grey exophenotype. Mr. Wilcock writes,

> "...a client of Dr. Newton's reports the Elders as all being hairless, with oval faces, high cheekbones, and smallish features— much like the appearance of certain types of extraterrestrials people have reported seeing. These people did have eyes like ours— not black ovals— but many extraterrestrial witness reports do feature this type of appearance."[18]

The implication of Mr. Wilcock's speculation, if true, is that the human "Interlife" experience of these subjects of Dr. Newton's would in fact be memories of a deceptive dimensional "virtual experience" in which the presumed Grey extraterrestrials in a dimension of the multiverse posed as advanced spiritual beings in the Spiritual dimensions.

On deeper analysis, however, one can reasonably conclude that Mr. Wilcock's speculation may be an unwarranted projection by Wilcock onto the data that Dr. Newton has collected about the reported Council of Elders. The full passage by Dr. Newton of an individual's soul memory of the Interlife that Mr. Wilcock is referring to is as follows:

---

[17] Newton, Destiny of Souls, Kindle Locations 619-634.
[18] Wilcock, The Synchronicity Key, pp. 194-195.

"One of the last requirements before embarkation for many souls is to go before the Council of Elders for the second time. While some of my subjects see the Council only once between lives, most see them right after death and just before rebirth. The spirit world is an environment personified by order and the Elders want to reinforce the significance of a soul's goals for the next life. Sometimes my clients tell me they return to their spirit group after this meeting to say goodbye while others say they leave immediately for reincarnation. The latter procedure was used by a subject who described this exit meeting in the following manner. "My guide, Magra, escorts me to a soft, white space which is like being in a cloud-filled enclosure. I see my committee of three waiting for me as usual. The middle Elder seems to have the most commanding energy. They all have oval faces, high cheekbones, no hair and smallish features. They seem to me to be sexless-or rather they appear to blend from male to female and back. I feel calm. The atmosphere is formal but not unfriendly. Each in turn asks me questions in a gentle way. The Elders are all-knowing about my entire span of lives but they are not as directive as one might think. They want my input to assess my motivations and the strength of my resolve towards working in new body. I am sure they have had a hand in the body choices I was given for the life to come because I feel they are skilled strategists in life selection. The committee wants me to honor my contract. They stress the benefits of persistence and holding to my values under adversity. I often give in too easily to anger and they remind me of this while reviewing my past actions and reactions towards events and people. The Elders and Magra give me inspiration, hope and encouragement to trust myself more in bad situations and not let things get out of hand. And then, as a final act to bolster my confidence when I am about to leave, they raise their arms and send a power bolt of positive energy into my mind to take with me."[19]

The full range of reported, replicable data about the intelligent civilization of souls that Dr. Michael Newton provides is incongruous with the Interlife being a deceptive virtual experience staged by Grey or other hyperdimensional extraterrestrials. If anything, Dr. Newton's data demonstrate that intelligent souls incarnate as Grey extraterrestrial exophenotypes in the various universes of the multiverse.

---

[19] Newton, Journey of Souls, Kindle Locations 3170-3176.

There is reliable evidence that advanced hyperdimensional extraterrestrials of the Grey exophenotype can undertake an operational role in the reincarnation cycle of intelligent souls from the Grey's platforms in the dimensional ecology of the universes of our multiverse. We will explore this evidence and how one species of Grey hyperdimensionals intervene in a soul's reincarnation cycle in the next chapter.

Parallels between Hyperdimensional Extraterrestrials and Souls in the Dimensional Ecology

Mary Rodwell, who has studied more than 6,000 cases of extraterrestrial contact, has discussed the many parallels that exist between the intervention of hyperdimensional extraterrestrials and the intervention of the intelligent civilizations of souls in the dimensional ecology of Earth humans. Ms. Rodwell writes,

> As we explore these potential realities, we find some [Extraterrestrial] encounters suggest that these intelligences may work or interact in dimensions of non-physical realities, such as the dimensions of soul. This encompasses Contact or communications with them, prior to our present physical incarnation, exampled by what Dr. Michael Newton describes in his book, *Destiny of Souls*, the life-between-life, soul state. Furthermore, my research suggests this interaction may also continue after we physically die. Testimony recounts that after our present human incarnation, we may remain involved with these Intelligences ... Interactions throughout what we would term time and space. I explore some of these extraordinary accounts that are consistent and seem to surface at the deeper level of hypnosis. They illustrate interactions that seem to have no limits in the space-time continuum. I have no way to personally verify any of this data.20

---

[20] Rodwell, "Extraterrestrials, human consciousness, and dimensions of soul," http://bit.ly/19TrVWY.

DIMENSIONAL ECOLOGY OF THE OMNIVERSE HYPOTHESIS: THE INTELLIGENT CIVILIZATIONS OF SOULS

We have reviewed representative excerpts and summaries by Dr. Michael Newton and his associates of more than 7,000 cases of replicable empirical evidence of the role of the intelligent civilizations of souls in the dimensional ecology of the Omniverse. These constitute *prima facie* evidence supportive of the dimensional ecology of the Omniverse hypothesis and include:

- Nature and creation of souls as holographic fragments of God
- Souls in the Interlife dimensions
- Soul groups in the Interlife dimensions
- Nature of Source (God) and dimensional ecology between Exopolitical and spiritual dimensions
- Parallels between hyperdimensional extraterrestrials and souls in the dimensional ecology
- Dimensional Ecology:
- Re-entry from the Exopolitical dimensions into the Interlife dimensions
- Life of a soul incarnated as an extraterrestrial
- Time travel, teleportation, and interdimensional portals between Exopolitical and Interlife dimensions
- Life creation in the universes of the Exopolitical dimensions (multiverse)
- Soul creation of galactic matter, stars, and planets in the Exopolitical dimensions
- Interdimensional travel to create or adjust planets in a universe
- Incarnations to alter time in the time-space hologram

- Soul interventions in dreams of individuals in the universes of the Exopolitical dimensions

We encourage you to read all of Michael Newton's original research to experience the full import of his discoveries about the nature of the dimensional ecology of the Omniverse.

Chapter 8

Hyperdimensional Civilizations' Role in Dimensional Ecology

As we have explored, replicable *prima facie* evidence supports the dimensional ecology of the Omniverse hypothesis by showing the extensive role that the intelligent civilizations of souls have both in the spiritual dimensions and in the universes of the Exopolitical dimensions (multiverse), in which souls incarnate in a diversity of lives. Souls also have an extensive role in the planning, creation, and husbandry of universes, galaxies, planets, and celestial bodies, as well as of living species of flora and fauna in the multiverse.

Role of Hyperdimensional Civilizations in Soul Reincarnation

There is *prima facie* eyewitness evidence that hyperdimensional civilizations based in our universe of the Exopolitical dimensions can play a central role in hosting souls on their spacecraft and facilitating soul incarnations into Earthly human bodies. This *prima facie* evidence demonstrates that human souls can incarnate into physical bodies as avatars, in order to have life experiences in the Exopolitical dimensions, in ways other than by teleporting directly from the interdimensional portal between the spiritual and Exopolitical dimensions to a location next to the womb of the mother of a future physical body.

One case reveals an important variation in the dominant method by which souls incarnate in our universe. In it, a soul was first introduced to its future Earthling human mother aboard a Grey hyperdimensional spacecraft when the mother was an eight-year-old child and had been teleported aboard the spacecraft. The mother and the soul, appearing as a "ball of light," played together under the observation of the Grey hyperdimensionals.

Years later, when the mother was married and pregnant, she was again teleported aboard a Grey hyperdimensional spacecraft and placed on a gurney for an "merging" procedure in which the soul of her future son was "merged" into the fetus of the mother's future son, then in her womb. The Grey hyperdimensionals told the mother details about the special role her son would play in the future in the time-space dimension on Earth.[1]

All these events took place in the multiverse, not in the Interlife dimensions of the Omniverse. This case indicates that, under specific circumstances, Grey hyperdimensionals have responsibility for enabling incarnation of souls in the Exopolitical dimensions that are normally relegated to "guides" and the life-selection and preparation process in the Interlife.

The case is that of Suzanne Hansen, Founding Director of UFOCUS NZ Research Network (New Zealand). Throughout her life, Hansen states, her experiences with the Grey hyper-dimensionals were "transformational, purposeful, and positive." Her account is corroborated by Mary Rodwell, RN, who led Hansen through hypnotic regression sessions designed to help

---

[1] Webre, "NZ UFO expert describes grey ET management of human souls & bodies aboard grey ET spacecraft," http://bit.ly/17Y7MSq; "NZ UFO expert: On a UFO, Grey ETs merged my son's soul into his body in my womb (Photos)," http://exm.nr/1eWlXJF.

her recover her memories of her life-long experiences around her son's soul aboard Grey ET spacecraft.

Mary Rodwell is a world leader in alien abduction counseling. She was featured in an SBS documentary alongside her son, a self-confessed UFO skeptic. Rodwell, a former nurse, midwife, and clinical hypnotherapist, founded ACERN in 1997 and has since investigated more than 1,600 cases of ET encounters worldwide. ACERN helps alien abduction victims deal with their traumas with the support of qualified medical doctors, psychologists, and therapists. In the documentary "My Mum Talks to Aliens," aired nationally in Australia, Rodwell and her son are followed on their journey of discovery as they travel the country in pursuit of solid evidence of an alien presence on Earth. "I believe these advanced intelligences have been visiting this planet since the dawn of time, and I believe they're here to assist evolution to another level," Ms. Rodwell said.

Suzanne Hansen has been a powerful and effective force in bringing about UFO and extraterrestrial disclosure in New Zealand, such as the disclosure of secret files by the Defense Force of New Zealand, which, on Wednesday, December 22, 2010, released more than 2,000 pages of reports by civilians, military personnel, and pilots, detailing unexplained aerial sightings.

In preparation for the report on this case, Hansen sent me a statement, accompanied by three drawings:

1   During this experience at age eight, I was initially taken to a large room in which I "played" with children of mixed species, using mind games to facilitate telepathic communications between us. The first [drawing] shows the large room ... I could "see" the objects being produced by our minds, with both my mind (thoughts) and my eyes. In another regression I did with Mary, I described being taught to project a hologram from my mind into the air in front of

me. You will notice a strange bed in this drawing, with a baby on it and a Grey beside it. In another regression session, Mary Rodwell regressed me to when I was 6 months old, where I could see myself lying on this small bed. I have nicknamed it the "lullaby bed" and a description of how and why this bed is used with human babies, and the technology involved, features in another chapter of my book.

2   was invited to accompany a Grey to a room where I was to meet another "child." The second drawing shows me as an eight year old, during my first "official" meeting with the ball of blue light, learning to relate to the soul of my future son, ... through play, chasing it around the room while the Greys "observed' from another area of the craft to see if we would be suited.

3   At age 28, I was taken onboard craft for a "merging" procedure, where the soul of my future son would enter my body, and that of my unborn child, for the first time. The third drawing shows the two entities (mixtures of Grey and another species) comforting me as a Grey begins a calming procedure to lower my metabolism prior to the soul of my son entering my body, and that of the unborn child.

I am very grateful to my husband, who is an artist, for doing draft drawings for my book. I sit with him for hours in order to get them as close as I can recall to the real thing.

HANSEN'S ACCOUNT OF SOUL INTERACTIONS ON BOARD GREY HYPERDIMENSIONAL CRAFT

In her interview with me, Hansen started by stating that at the age of eight years she had been taken by Grey extraterrestrials on board a Grey spacecraft and told that she would be meeting someone who would be "very important" to her in her life. She states that she had had UFO experiences with her family throughout her life, so that this particular experience was not that strange to her.

As an adult, Hansen began to have flashback experiences of this event and asked Mary Rodwell to do hypnotic regressions to fill in the gaps in her memories. During the regressions with Rodwell, Hansen recovered memories of being taken aboard a Grey ET craft. She played with a group of Grey and transgenic (hybrid) children. All the play was telepathic. She recalls crafting holographic shapes in the air with other children and playing with the shapes as toys.

On one occasion when she was eight years old, Hansen recalls, a Grey hyperdimensional who she felt was holding something back approached her. The hyperdimensional introduced her to a "Ball of Light" on the ship. Upon seeing the Ball of Light, she felt confused about the Light. The hyperdimensional asked her to look into his eyes. Hansen said that she felt three distinct streams of information coming from him.

The hyperdimensional first communicated that the Ball of Light was a friend, whom she would like to come and visit. Second, the hyperdimensional communicated a full packet of visual information to her about the life of a child who was the Ball of Light. Hansen recalls turning cartwheels on the spacecraft room floor as she tried to relate to the Ball of Light. Finally, the hyperdimensional communicated to her that she and the Ball of Light had known each other at the soul level for a long time in previous incarnations.

During the years after this experience, Hansen states she became familiar with the soul state. She became able to communicate with persons who were in the afterlife (or Interlife) in the soul state.

Hansen recovered the memories that, after her marriage and while pregnant with her son, she was taken aboard a Grey hyperdimensional craft. On the craft, she recalls being escorted by a Grey hyperdimensional (who was familiar because of his "frequency output") down a corridor into a small room where she saw come unfamiliar medical equipment. At that point, a shorter Grey hyperdimensional, perhaps four feet tall, came into the room, with the archetype of a "short Grey" and different in appearance from the other Greys in the room.

Dimensional Ecology: The Soul-Merging Procedure

It was at this point that a "merging procedure" began. First, Hansen recalls, the Grey hyperdimensional needed to retrieve some of her earlier memories of interacting with the Ball of Light. The hyperdimensional placed one hand over her forehead, and Ms. Hansen saw visual memories of her communicating as an eight-year-old child with a blue Ball of Light on a Grey spacecraft. She became tearful and emotional as she recovered these memories. She was then told that her son's soul had been enhanced as a Ball of Light, whose signature was more powerful and could harm her if a procedure were not undertaken to lower her metabolism.

At that point, Ms. Hansen recalls, about 18 Grey hyperdimensional children entered the room to observe the soul-merging procedure. She recalls they were chattering away, telepathically. The hyperdimensional told her these children, who were on the way to growing up, were called "Watchers." They appeared to be working with the hyperdimensional on a specific task.

Hansen recalls that at that point her metabolism was lowered. She was barely breathing and her body temperature was very low. Her voice was slow and quiet. The blue Ball of Light then entered her body and the body of the fetus of her son in her womb. She recalls that the metabolism of her body was then raised. She was then given information about her son's future, and about how the child would be educated both on board the

Grey hyperdimensional ship and in the human dimension on Earth. Her son was to receive this dual education for the benefit of mankind, she was told.

Hansen stated that her son as of 2011 had no memory of being aboard the Grey hyperdimensional craft. The Grey hyperdimensionals have told her that when her son is in his mid-30s, he will recover his memories (he was in 2011 in his mid-20s). She has discussed this history of their prior interaction on the Grey hyperdimensional ship with her son. He now has abilities such as psychic capacity and precognition.

Who Are the Greys and What Is Their Mission?

Hansen states that she knows there are many species of extraterrestrials now visiting Earth. Some of these species are working together. Rodwell, for example, has stated that one expert opinion holds there are more than 150 species of Grey hyperdimensionals known to exist.

It is a veritable minefield, Hansen says, to understand fully what the mission of the Greys might be, because some people's experiences with Grey hyperdimensionals are fearful; some people's experiences are transformational, etc. Her experiences with the specific species of Grey hyperdimensionals she encountered are purposeful and transformational, although she has had to work through a period of personal fear about the experiences.

Hansen says that her understanding is that souls are universal and can chose to incarnate into a variety of species on different planets. She says that a soul is a spiritual entity, and all of its incarnations appear designed for it to learn moral or evolutionary lessons. She states that she perceives that the Grey hyperdimensionals cannot "command the human race." There is too much chaos on this planet (Earth). The Grey hyperdimensionals' way appears to be to work with souls who are ready to incarnate on a planet and approach its people with love and education. When I asked Hansen if she knows how souls are created and how it is that souls come to be on Grey hyperdimensional craft, she demurs and states she does not know.

Mary Rodwell's Corroborative Interview

Mary Rodwell's interview with me corroborates the events of Hansen's recollection of her experiences on the Grey ET craft. Rodwell's case materials from her own work as an abduction researcher shed light on how human souls apparently have access to Grey hyperdimensional ships in the form of Balls of Light.

Rodwell mentions a 15-year-old boy who saw two Balls of Light accompanying him during an encounter with Grey hyperdimensionals. The boy stated that the Balls of Light were the souls of "Dad and Grandfather," who were at the scene to encourage the boy not to be frightened and to let him know that everything was "OK." Rodwell comments that this case illustrates that Grey hyperdimensionals work with human souls in a very tangible way, although they appear to be intelligences that are different from souls.

Rodwell says her abduction research has led her to conclude that the job of this specific species of Grey hyperdimensionals includes:

1. Gathering genetic information and materials on humans;
2. Carrying out healing procedures on humans;
3. Cooperation between interdimensional intelligences, as evidenced by the cooperation between souls and Grey hyperdimensionals;
4. Deepening of compassion and love, healing ability, and telepathic transfer of information by souls incarnated as Earthling humans.

Of the 1,600 abduction cases Rodwell has investigated, approximately 30 percent involved some level of trauma to the human abductee. The remaining 70 percent of cases do not involve trauma.

The medical and health inspections of humans by the Grey hyperdimensionals appear to be for levels of pollution, toxins, and other harmful materials that relate to the planet's problems.

Souls Common to Extraterrestrials and Earthlings

In 2011 I developed a hypothesis like that of researchers Suzanne Hansen and Mary Rodwell that integrates extraterrestrials and the human soul and is based on empirical data. The hypothesis corroborates and validates their data and findings, as well as the plausibility and veracity of the case study reported in this article.[2]

Based on extensive research findings in Exopolitics, parapsychology, reincarnation studies, and hypnotic regression, this hypothesis postulates that an ecology of spiritual dimensions mirrors the parallel dimensions that science is now researching and that include our known physical universe.

Empirical research on the Interlife demonstrates that a civilization of intelligent, evolving souls is apparently created in the spiritual dimensions. These souls in turn evolve and incarnate into various human or other forms on diverse planets in the known physical universe and other planets, for purposes of gaining moral lessons. This hypothesis is supported by replicated data from hypnotic regression of multiple subjects' Interlife memories.

---

[2] The content in this section originally appeared in Webre, "NZ UFO Expert describes grey ET management of human souls & bodies aboard grey ET spacecraft," http://bit.ly/17Y7MSq.

The typology of intelligent civilizations

The dimensional ecology hypothesis of the multiverse includes a new typology of extraterrestrial civilizations, based on empirical evidence that confirms the existence of the following typologies of extraterrestrial civilizations and governance bodies:

A *Solar-system civilizations*: Planetary civilizations in our own solar system, such as the intelligent human civilization living under the surface of Mars that reportedly enjoys a strategic relationship with the United States government.

B *Deep-space civilizations*: Intelligent extraterrestrial civilizations based on a planet, solar system, or space station in our galaxy or in some other location in this known physical universe. An example includes the ten to a hundred intelligent extraterrestrial civilizations that, according to a NATO report, have been visiting Earth for many centuries.

C *Hyperdimensional civilizations*: Intelligent civilizations that are based in dimensions or universes parallel to our own and that may use technologically advanced physical form and/or transport when entering our known universe.

D *Extraterrestrial governance authorities*: Legally constituted extraterrestrial governance authorities with jurisdiction over a defined territory, such as the Milky Way Galactic Federation, which has been empirically located in replicable research.

A dimensional ecology

The dimensional ecology hypothesis, which concludes that we exist in a dimensional ecology consisting of an Exopolitics dimension and a spiritual dimension, was developed after extensive review of research in extraterrestrial studies, afterlife studies, reincarnation studies, and the results of hypnotic regression to recover Interlife memories. The scientific data support the existence of an Exopolitical dimension, which has been traditionally been called "the physical universe" and which consists of parallel universes and dimensions, including our own known universe. This dimension is called Exopolitical because souls from the spiritual dimension incarnate into the Exopolitical dimension for experiences as physical beings on diverse planets, hence as diverse "extraterrestrials." In analyzing data from many cases involving hypnotic regression from Interlife memories, I extended the hypothesis that a community of souls is developed in the spiritual domain in order to propose that the community of souls is reincarnating in complex patterns as extraterrestrials and earthlings. Hence souls are common to extraterrestrials and earthling humans.

Dimensional Ecology: Relationships Between Grey Hyperdimensionals and the Intelligent Civilizations of souls
Both the Suzanne Hansen case and the cases developed by Michael Newton support the dimensional ecology of the Omniverse hypothesis. The Hansen case and other cases of operating relationships between the intelligent civilizations of souls and specific species of Grey hyperdimensionals do not contradict the replicable findings from hypnotic regression of soul memories of the Interlife of the 7,000 cases developed by Newton.

The Hansen case demonstrates that souls and specific Grey hyperdimensionals navigate the dimensional ecology of the multiverse as interdimensional entities in cooperative relationships, such as mutual assistance in education of a future parent of an incarnating soul and in the actual incarnation process. Thus some souls, such as those of Suzanne Hansen's future son and those of the reported father and grandfather of the 15-year-old boy who accompanied him during an encounter with Grey hyperdimensionals, appear to be part of a structured program of cooperation among the intelligent civilizations of souls and this specific species of Grey hyperdimensionals for specialized intervention in specific types of soul incarnations or educational encounters with Grey hyperdimensionals in the Exopolitical dimension.

At least one organization researching the role of Grey hyperdimensional extraterrestrials has preliminarily concluded that a specific species of Grey extraterrestrials may be carrying out an extensive intervention in the dimensional ecology of the Earth human reincarnation cycle.[3] Earth humans are at the beginnings of our collective understanding of the full dimensional factors that may be affecting the human soul's reincarnation cycle. This is a priority area for our further research, understanding and species cosmic growth.

---

[3] Joseph Montaldo, Research discussion on "UFO Undercover", International Community for Alien Research, February 12, 2014, http://bit.ly/1aEN6v7

CHAPTER 9
THE SCIENCE OF THE OMNIVERSE

There are at least ten implications for a positive human future that can be drawn from this research into the science of the Omniverse for further exploration of the dimensional ecology of the Omniverse hypothesis.

1. *PRIMA FACIE* EMPIRICAL EVIDENCE SUPPORTS THE DIMENSIONAL ECOLOGY OF THE OMNIVERSE HYPOTHESIS

A reasonable observer can now conclude that *prima facie* empirical evidence supports the dimensional ecology of the Omniverse hypothesis. This hypothesis holds that we Earthlings live in a dimensional ecology of intelligent life that encompasses intelligent civilizations based in parallel dimensions and universes [the Exopolitical dimensions or the multiverse], and souls, spiritual beings, and Source (God) in the spiritual dimensions. Together, the Exopolitical dimensions and the spiritual dimensions form the Omniverse. The totality of the spiritual dimensions (souls, spiritual beings and God itself) functions as the Source of the universes of the Exopolitical dimensions.

2. EXOPOLITICS AND PARAPSYCHOLOGY

Exopolitics, the science of relations among intelligent civilizations[1], and parapsychology, the science of psi consciousness, telepathy, reincarnation, the soul, the Interlife, and Source (God),[2] are among the proper scientific disciplines for exploring and mapping the dimensional ecology of the Omniverse.

It is time that Exopolitics and parapsychology be admitted to the accepted canon of sciences and openly taught at all levels of education, elementary, secondary, and post-secondary. The state censorship barriers of official ridicule and denial that were constructed in 1953 by the US Central Intelligence Agency in its Robertson Panel Durant Report against any civil society discourse, publication, research, or public knowledge of a populated, organized Omniverse must now be forever deconstructed and discarded. The governments and intelligence communities, the financial and political elites that control these do not have the spiritual, moral, or legal authority to suppress research and access to the Omniverse, including the Exopolitical dimensions, the Interlife, the spiritual dimensions, and Source (God).[3]

3. DIMENSIONALITY IS A KEY DESIGN CRITERION OF THE OMNIVERSE

Dimensionality, the ability of intelligence to organize itself via dimensions (discrete bands of conscious energy), appears to be a key criterion by which the Omniverse is designed, both in the spiritual dimensions and in the Exopolitical dimensions. Intelligent civilizations, in both the spiritual and the Exopolitical dimensions,

---

[1] Webre, Exopolitics: Politics, Government and Law in the Universe.

[2] For basic information on aspects of parapsychology, see the works of Chris Carter, Russell Targ, and Michael Newton.

[3] Report of Scientific Advisory Panel on Unidentified Flying Objects Convened by Office of Scientific Intelligence, CIA, January 14-18, 1953 (the Durant report of the Robertson Panel proceedings), http://bit.ly/15ry5Na.

can be typed by the dimension in which they are based. The principle of dimensionality underlies the dimension-based typology of extraterrestrial and interdimensional civilizations, for example.

4. Human Soul Consciousness and Development

The consciousness and developmental level of souls incarnating on Earth and our collective ability to comprehend the dimensional ecology of the multiverse is both a bottleneck and a key to the future evolution of Earthling humans in the society of organized intelligent life in the Exopolitical dimensions.

Commenting on his replicable findings from hypnotic regression of soul memories of the Interlife about how souls are distributed by level of development (beginner, intermediate, advanced souls), Michael Newton writes, "I believe almost three-quarters of all souls who inhabit human bodies on Earth today are still in the early stages of development. I know this is a grossly discouraging statement because it means most of the human population is operating at the lower end of their training … . Nevertheless, I see the possibility that we may have only a few hundred thousand people [souls] on Earth at level V [Highly Advanced]."[4]

5. Positive Transformation of Power, Economic, and Social Structures on Earth

The long-term positive transformation of power, economic, and social structures on Earth may depend on "Soul Power." Significant soul development during an incarnation can occur as breakthroughs in how science-based knowledge and information about the dimensional ecology of the Omniverse is:

---

[4] Newton, Journey of Souls, p. 123.

A   seriously internalized by the more advanced and intermediate souls incarnating on Earth, and
B   presented to the approximately three-quarters of the Earth's population who incarnate beginner's souls (assuming that Newton's findings are empirically correct) in
C   context in which the dimensional ecology itself and hyperdimensional civilizations within the dimensional ecology are supporting an expansion of Earthling human consciousness.

Carl Johan Calleman has hypothesized that, from October 28, 2011, onward, the singularity or interdimensional portal at the center of our galaxy (the galactic center "black hole") has been emanating constant "enlightened universal Unity consciousness." According to Calleman, the singularity or interdimensional portal modulates this universal Unity consciousness energy wave as well. This is, by any standard, exceedingly good scientific news.

Universal, alternating energy-wave movements have been a feature of our universe, says Calleman, since the Big Bang, and it is these wave movements that have shaped the nature of consciousness in our universe over the past 18 billion years.

In my interview with him, Calleman stated, "The universal alternating energy-wave movements [have ended], and Earth is set on a gradual setting of a potential to reach advanced utopian planet status, a virtual 'Garden of Eden'."[5] This milestone marks the beginning of final preparations for the entry of Earth into a long-prophesized Golden Age of Utopia, or "positive timeline."

What, you may ask, does this universal wave of unity consciousness have to do with you or your reality? In a word, everything.

The wave of unity consciousness, like all universal energy-wave emanations, creates the "meme" or story content of our

---

[5]   Webre, "GOOD NEWS! Universe singularity now emanating pre-wave energy for 'enlightened unity consciousness'," http://bit.ly/1avePQ0.

personal and collective reality. Calleman's discoveries suggest that our highest thoughts and "memes" may in fact be sourced from the energy waves of the universal singularity, mediated through the dimensional ecology inside our universe via the galactic singularity, the black hole at the center of our galaxy that the Mayans called Hunab Ku.

The singularity of our solar system is Sol, our sun, a dimensional portal to other galaxies, according to physicist Nassim Haramein. These universal singularity energy waves serve as carrier waves for "the universe's mind and spirit software." That is, they are the way that an intentional universe (as created and maintained by a collective, conscious Source) lets us know what it intends for consciousness in our entire universe.

As a practical matter, does it matter if you are tuned into and committed to the universal energy wave of unity consciousness? Yes, the universal energy waves we humans tune into and commit to determine our planetary status and how we reach our potential as a planet.

Will Earth be a dystopia or a utopia? The good news is that our galactic singularity is now emanating "utopia" consciousness waves on its alternating energy carrier waves. The more we humans can individually and collectively tune into the universal energy wave of "enlightened universal consciousness, and can commit to achieving nondualistic consciousness, then the stronger the "positive timeline" becomes, and the more rapidly we will achieve our potential as a planet to reach advanced utopian planet status.

The coming global "shift"

Extraterrestrial researcher Mary Rodwell shared details of a coming global shift that she has discovered in her investigations of conscious interaction with extraterrestrials, including at the Soul level. The coming shift is described as including an awakening and transformation of human consciousness, along with dramatic changes in the way human society functions on the Earth.[6] Some researchers state that this global shift of consciousness is a long-term dimensional project conducted by Grey hyperdimensionals and may be limited to the contactees and abductees of the Greys.

6. Comprehending Extraterrestrial/Hyperdimensional and Exopolitical-related Phenomena and Information

The dimensional ecology of the Omniverse hypothesis and the dimension-based typology of intelligent civilizations in the Exopolitical dimensions can facilitate and accelerate the comprehension of UFO and extraterrestrial-related data and information by the Earthling human public, and by scientific, governmental, educational, and media organizations. The dimensional ecology of the Omniverse hypothesis and the dimension-based typology of intelligent civilizations in the Exopolitical dimensions can thus accelerate the pace of extraterrestrial disclosure, including disclosure by these intelligent civilizations themselves.

7. Comprehension of the Spiritual Dimension: Nature of Soul, Interlife, God

The dimensional ecology of the Omniverse hypothesis and the science-based study of the spiritual dimension can facilitate and accelerate the comprehension of true versions of the basic concepts of reality, such as soul, Interlife and Source (God), by the Earthling human public, and by scientific, governmental, educational, and media organizations.

Humanity is at the threshold of understanding the nature of the dimensional ecology of the Omniverse in which humanity is

---

[6] Webre, "Mary Rodwell – ETs, souls, the new humans. and a coming global shift," http://bit.ly/17tBY7B.

based and operates. Open questions, such as whether more than one Omniverse and one Source (God) exist, can be addressed. The dimensional ecology of the Omniverse hypothesis and the science-based study of the spiritual dimension can thus accelerate the pace of conscious evolution of Earthling humanity and of the population of souls incarnating on Earth.

8. The Centrality of the Intelligent Civilizations of Souls to the Universes of the Exopolitical Dimensions

The science-based study of the dimensional ecology of the Exopolitical and spiritual dimensions of the Omniverse reveals the centrality of the intelligent civilizations of souls in creation, maintenance, design, and ultimately incarnation in the universes of the Exopolitical dimensions.

A common denominator to all the dimensions of the Omniverse is the soul, which is a holographic fragment of the Creator Source (God). The human soul inhabits the spiritual dimensions between physical lives and inhabits the Exopolitical dimensions during lives (and may lead several Exopolitical lives simultaneously). Exopolitical lives may be as a human on Earth, or an intelligent being on another planet or in another universe or dimension. There may be a multiplicity of types of intelligent souls that our science will discover.

One species of Grey hyperdimensionals appears to be significant in assisting soul incarnation, at least of the Third Wave Indigos, born and emerging in the 21st century. There are some reports of "soul capture and mistreatment" by some Grey hyperdimensionals. Living humans report meeting the souls of their "deceased" relatives aboard Grey hyperdimensional spacecraft. There is preliminary contactee witness evidence that a specific species of Grey hyperdimensionals may have an extensive role in the reincarnation cycle of Earth humanity's souls. This is a vital area for future research and for humanity's understanding of the dimensional ecology of its incarnating on Earth.

9. Education about the Soul, Interlife, and Source (God)

Humanity now is being misinformed about the true nature of the soul, the Interlife, the mechanisms of reincarnation, and ultimately of Source (God). Religions are a large source of erroneous information about these realities. This misinformation is based on texts and ancient religious belief systems that are not scientifically correct and are yet considered sacred, as a matter of faith.

Earth religions have resulted from manipulatory extraterrestrial intervention, such as the Anunnaki extraterrestrial intervention on Earth starting approximately 280,000 years ago.[7] Academic science prohibits teaching the reality of life after death, even though empirical evidence supports this hypothesis. Therefore, modern science is also a source of erroneous information about the soul, the Interlife, and Source (God). Cultural and religious conflicts result from the ignorance imposed by religion and an antiquated scientific canon concerning any such spiritual realities. The science of spirit can demonstrate the true nature of the human soul and its spiritual and exopolitical dimensions.

10. Science and Spirituality

The dimensional ecology of the Omniverse hypothesis is itself a matrix for recognition and classification of ongoing and new research into intelligent civilizations in the Exopolitical and

---

[7] Tellinger and Heine, Temples of the African Gods, 2009. See my film "Occupy Adam's Calendar - Part I Extraterrestrial Genetic Manipulation: Geneticist William Brown," http://bit.ly/W9o4jH.

spiritual dimensions. This hypothesis represents a coming together of science and spirituality that will return science to the proper study and understanding of the human soul, and a return to the supremacy of understanding over ignorance. To support and encourage this understanding is our duty as informed, aware human souls.

Sources and Resources

Albo, Frank, and Manitoba Culture, Heritage and Citizenship. *The Cipher Golden Boy: Occult Symbolism in the Manitoba Legislative Building.* Winnipeg, Manitoba: Manitoba Culture, Heritage and Citizenship, 2010.

Allen, Rayelan. *The Obergon Chronicles.* Amazon.com, 2006. http://amzn.to/KmCA4z.

Amoroso, Richard L., and Elizabeth A. Rauscher. *The Holographic Anthropic Multiverse: Formalizing the Complex Geometry of Reality.* Oakland, CA: Noetic Advanced Studies Institute, Technic Research Laboratories, 2009. http://bit.ly/1eX5N12.

Ann, Kristen. *Exploring Sacred Space.* Vast Quest Ventures, 2011. Amazon.com, http://amzn.to/1kNEtY9.

Anthony, Sebastian. "Astronomers estimate 100 billion habitable Earth-like planets in the Milky Way, 50 sextillion in the universe." *http://bit.ly/13UiMgU.*

Associated Press. "United Nations hears case for UFOs" [article about Eric Gairy, Prime Minister of Grenada], October 15th, 1978.

Australian Close Encounter Research Network (ACERN). http://bit.ly/18DVPed.

Barrett, Sonia. *The Holographic Canvas.* No. Hollywood, CA: Timeline Publishing, 2007.

Basiago, Andrew D. "Three principal Exophenotypes of humanoid beings on Mars." http://bit.ly/1cahkeq.

———. "Hoagland affirms belief that no evidence exists Mars is inhabited." http://bit.ly/16yOv9f.

———. Project Pegasus. http://on.fb.me/19pjqmo.

———. *The Discovery of Life on Mars*. Vancouver, WA: Mars Anomaly Research Society, 2008. *http://bit.ly/1837faW* [this link will download the PDF file to your computer].

Bateman, Wesley H. *Through Alien Eyes*. Sedona, AZ: Light Technology Publishing, 2000.

Belitsos, Byron. "Introducing the Urantia Book: A Revelation of Love and Cosmic Evolution." In *The Adventure of Being Human* (San Rafael, CA: Origin Press, 2012).

Berliner, Don, and Whitley Streiber (compiler). "International Agreements and Resolutions – United Nations," in *UFO Briefing Document: The Best Available Evidence*. Dell, 2000.

Berman, Bob, and Robert Lanza. *Biocentrism: How Life and Consciousness Are the Keys to Understanding the True Nature of the Universe*. BenBella Books, Kindle Edition, 2010.

Bertell, Rosalie. *Planet Earth: The Latest Weapon of War*. Montreal: Black Rose Books, 2001.

Blazquez, Francisca, and Joan Lluis Montane. *El Paraiso de la Nueva Luz*. Madrid, Spain: Mandala Ediciones, 2010. http://bit.ly/1cJfco9.

Blazquez, Francisca. *Dimensionalismo*. Madrid, Spain: Museo Casa de la Moneda, 2007.

Bourdais, Gildas. *COMETA: The French Report on UFOs and Defense: A Summary*. Chicago, IL: Center for UFO Studies (CUFOS), 1999.

Bromberg, Facundo. "On Goswami's monistic idealism worldview." *www.cs.iastate.edu/~bromberg/BrombergPhyl465FinalPaper.pdf*.

Brown, Courtney, Ph.D. *Cosmic Explorers*. Dutton, 1999.

———.*Cosmic Voyage*. Dutton, 1996.

———. *Remote Viewing: The Science and Theory of Non-Physical Perception*. Farsight Press, 2005. http://bit.ly/1iB4cNb.

Calleman, Carl Johan, Ph.D. *The Mayan Calendar and the Transformation of Consciousness*. Rochester, VT: Bear & Co., 2004.

———. *The Purposeful Universe*. Santa Fe, NM: Bear & Co., 2009.

Cameron, Grant. "UFO studies done and proposed by the Carter administration." *www.presidentialufo.com*, 2005.

Cameron, Tilde, and Tina Fiorde. *A Book of Insight*. Copquitlam, BC: Stone Circle Publishing, 2010.

Carlsberg, Kim. *Beyond My Wildest Dreams*. Rochester, VT: Bear & Co., 1995.

———. *The Art of Close Encounters*. Sedona, AZ: Close Encounters Publishing, 2010. http://bit.ly/1ahRoNP.

Carroll, Judy. *Human by Day, Zeta by Night: A Dramatic Account of Zetas Incarnating as Humans*. Columbus, NC: Wildflower Press, 2011.

Carrol, Judy, and Helene Kaye. *The Zeta Message: Connecting All Beings in Oneness*. Columbus, NC: Wildflower Press, 2011.

Carter, Chris. *Science and Psychic Phenomena*. Rochester, VT: Inner Traditions, 2007.

Chalker, Bill. *Hair of the Alien: DNA and Other Forensic Evidence of Alien Abduction*. Pocket Books, 2005.

Chevalier, Andre. *I Speak to the Dead*. Montreal: Amazon.com, 2010. http://amzn.to/1dUCBGU.

Clow, Barbara Hand. *Awakening the Planetary Mind: Beyond the Trauma of the Past to a New Era of Creativity*. Rochester, VT: Bear & Co., 2011.

———. *Catastrophobia: The Truth Behind Earth Changes in the Coming Age of Light*. Rochester, VT: Bear & Co., 2001.

———. *The Mayan Code: Time Acceleration and Awakening the World Mind*. Rochester, VT: Bear & Co., 2007.

———. *The Mind Chronicles: A Visionary Guide into Past Lives*. Rochester, VT: Bear & Co, 2006.

Clow, Barbara Hand, with Gerry Clow. *Alchemy of Nine Dimensions*. Charlottesville, VA: Hampton Roads, 2004.

———. *The Pleiadian Agenda: A New Cosmology for the Age of Light*. Santa Fe, NM: Bear & Co., 1995.

Cockell, C.S. "Liberty and the limits to the extraterrestrial state." *J. Brit. Interplanetary Soc.*, 62 (2009), 139-157. http://bit.ly/1crfovd .

Colli, Janet Elizabeth. Ph.D. *Sacred Encounters: Spiritual Awakenings during Close Encounters*. Bloomington, IN: Xlibiris, 2004.

Colombo, John Robert. *Extraordinary Experiences*. Toronto: Hounslow Press, 1999.

Cook, Nick *The Hunt for Zero Point*. Broadway Books, 2002.

Cooper, L. Gordon. "Astronaut Gordon Cooper addressing UN panel discussion on UFOs and ETs, 1985." *UFO Universe*, vol. 1, no. 3 (November 1988), at *www.ufoevidence.org*/documents/ doc961.htm.

Corak, Vedia Bulent (Onsu). *The Knowledge Book*. Istanbul, Turkey: World Brotherhood Union Mevlana Supreme Foundation, 1995.

Cori, Patricia. *Before We Leave You*. Berkeley, CA: North Atlantic Books, 2011.

———. *No More Secrets, No More Lies*. Berkeley, CA: North Atlantic Books, 2008.

Corso, Philip J. *The Day After Roswell*. Pocket Books, 1997.

DailyGalaxy.com. "500 Billion—A Universe of Galaxies: Some Older than Milky Way." *http://bit.ly/1grvKpP*.

David, Leonard. "UFO Group Demands Congressional Hearing." www.space.com, May 9th, 2001.

Deane, Ashayana. *Voyagers: The Sleeping Abductees*. Columbus, NC: Wildfire Press, 2001.

Deardorff, James, Bernard Haisch, Bruce Maccabee, and Hal E. Puthoff. "Inflation-theory implications for extraterrestrial visitation." *J. Brit. Interplanetary Soc.*, 58 (2005), 43-50.

Dewhurst, Richard J. *The Ancient Giants who Ruled America*. Rochester, VT: Bear & Co., 2014.

Dickinson, Terence, and Adolf Schaller. *Extraterrestrials: A Field Guide for Earthlings*. Buffalo, NY: Camden House, Firefly Books, 1994.

Disclosure Project. *Disclosure, Military and Government Witnesses Reveal the Greatest Secrets in Modern History*. Crozet, VA: The Disclosure Project, 2001.

Dolan, Richard M. *UFOs and the National Security State: Chronology of a Coverup, 1941-1973*. Amazon Kindle, 2002.

Dolan, Richard M., and Linda Moulton Howe. *UFOs and the National Security State, Vol. 2: The Cover-Up Exposed, 1973-1991*. Amazon Kindle 2010.

Dolan, Richard M., Bryce Zabel, and Jim Marrs. A.D. *After Disclosure: When the Government Finally Reveals the Truth About Alien Contact.* Amazon Kindle, 2012.

Dvorsky, George. "How to measure the power of alien civilizations using the Kardashev scale," February 25, 2013. http://bit.ly/J3G0Yy.

Dyer, M.G. "Quantum physics and consciousness, creativity, computers: A commentary on Goswami's quantum-based theory of consciousness and free will." *J. Mind and Behavior*, 15(1994), 265-290.

Eisenhower, Jean. *Rattlesnake Fire.* Silver City, NM: ParadigmSalon Publishing, 2010.

Elder, Paul. *Eyes of An Angel.* Charlottesville, VA: Hampton Roads, 2005.

Elkins, Don, Carla Rueckert, and James Allen McCarthy. *The Ra Material: An Ancient Astronaut Speaks (The Law of One , Nos. 1 - IV)*, 1984. Cassiopedia, http://bit.ly/GJZMaF.

Ellegion, Michel, and Aurora Light. *Prepare for the Landings.* Scottsdale, AZ: Vortex, 2008.

Ellis, George F. R. "Does the multiverse really exist?" *Scientific American*, July 19, 2011. *http://www.scientificamerican.com/article.cfm?id=does-the-multiverse-really-exist.*

Essene, Virginia, and Sheldon Nidle. *You are Becoming A Galactic Human.* Santa Clara, CA: Spiritual Education Endeavors, 1994.

Eure, Robert Frank. *The Mysterious Vistors.* Bloomington, IN: Xlibris, 2004.

Fernandes, Fernando, Joaquim Fernandes, and Raul Berenguel. Andrew D. Basiago and Eva M. Thomson, eds. *The Fatima Trilogy.* San Antonio, Texas: Anomalist Books, 3 vols., 2008.

Flem-Ath, Rand, and Colin Wilson. *The Atlantis Blueprint.* Time-Warner, 2000.

Fulham, Stanley A. *Challenges of Change.* Winnipeg, Manitoba: Amisk Enterprises, 2010.

Fontana, David. *Is There An Afterlife? A Comprehensive Overview of the Evidence.* Blue Ridge Summit, PA: NBN Books, 2005. Amazon, http://amzn.to/1f3jY2K.

— — —. *Life Beyond Life: What Should We Expect?* London: Watkins Publishing, http://amzn.to/1ebVVN4

Garland, Joe. "The Kardashev scale – Type I, II, III, IV & V civilization." October 8, 2013. http://bit.ly/1bNRJBm.

Garriga, Jaume, and Alexander Vilenkin. "Many worlds in one." *Phys. Rev.*, vol. D64, no. 043511 (July 26, 2001). Available online at arXiv.org/abs/gr-qc/010 2010.

Gibson, Mitchell Earl, M.D. *Your Immortal Body of Light*. Forest Hill, CA: Reality Entertainment, 2006.

Good, Timothy. *Alien Contact*. William Morrow, 1993.

Goswami, Amit. "The idealistic interpretation of quantum mechanics." *Phys. Essays*, 2 (1989), 385-400.

———. "Consciousness in quantum physics and the mind-body problem." *J. Mind and Behavior*, 11, 1 (Winter 1990), 75-96.

———. *The Self-Aware Universe: How Consciousness Creates the Material World:*. Tarcher/Putnam, 1993.

———. "Monistic idealism may provide better ontology for cognitive science: A reply to Dyer." *J. Mind and Behavior*, 16, 2 (Spring 1995), 135-150.

Greene, Brian. *The Hidden Reality: Parallel Universes and the Hidden Laws of the Cosmos*. Knopf, 2011.

Greene, Brian, et al. "What is life like in other parts of the multiverse?" [Video] *Scientific American*, November 21, 2011. http://www.scientificamerican.com/article.cfm?id=fabric-of-the-cosmos-multiverse.

Greer, Steven M., MD. *Disclosure: Military and Government Witnesses Reveal the Greatest Secrets in Modern History*. Crozet, VA: Disclosure Project, 2001. Amazon.com, http://amzn.to/1eKVlaE.

———. *Extraterrestrial Contact: The Evidence and Implications*. Amazon Kindle, 2013.

———. *Hidden Truth: Forbidden Knowledge*. Amazon Kindle, 2013.

Haitch, Elizabeth. *Initiation*. Santa Fe, NM: Aurora Press, 2000. Originally published, 1960.

Hardcastle, Rebecca, Ph.D. *Exoconsciousness: Your 21st Century Mind*. Amazon Kindle, 2008.

Harper, John Jay. *Tranceformers: Shamans of the 21st Century*. Forest Hill, CA: Reality Entertainment, 2006.

Harris, Fred, and Byron Belitsos, eds. *The Center Within*. Novato, CA: Origin Press, 1998.

Harris, Paola Leopizzi. *Connecting the Dots: Making Sense of the UFO Phenomenon*. Bloomington, IN: Author House, Amazon Kindle 2005.

———. *Exopolitics: All the Above*. Bloomington, IN: Author House, 2009.

———. *Exopolitics: How Does One Speak To A Ball Of Light?* Bloomington, IN: Author House 2007.

Hawkins, David R., M.D., Ph.D. *Power vs Force*. Carlsbad, CA: Hay House, 2002.

Haze, Xaviant. *Aliens in Ancient Egypt*. Rochester, VT: Bear & Co., 2013.

Hellyer, Paul. "Extraterrestrials want to help mankind". Russian Television, *http://bit.ly/1hYqxWn*.

———. *Light at the End of the Tunnel*. Bloomington, IN: Author House, 2010.

Helmenstine, Anne Marie, Ph.D. "Scientific hypothesis, theory, law definitions, learn the language of science." *http://bit.ly/bV90Rj*.

Horowitz, Leonard. *Walk on Water*. Sandpoint, ID: Tetrahedron, 2006.

Huneeus, Antonio. "Statement to the Second Symposium on Extraterrestrial Intelligence and the Human Future, UN Society for Enlightenment and Transformation (SEAT)," reprinted in SEAT Newsletter, December 1993.

Hurtak, J.J. *The Keys of Enoch*. Los Gatos, CA: The Academy for Future Science, 1977.

Huyghe, Patrick, and Dennis Stacy. *The Field Guide to Extraterrestrials: A Complete Overview of Alien Life Forms Based on Actual Accounts and Sightings*. Avon Books, 1996.

Hynek, J. Allen, "Dr. J. Allen Hynek Speaking at the United Nations," November 27th, 1978, at *www.ufoevidence.org/documents/doc757.htm*.

———. *The UFO Experience: A Scientific Inquiry*. Marlowe & Co., 1998.

Hynek, J. Allen, and Jacques Vallée. *The Edge of Reality: A Progress Report on the Unidentified Flying Objects* (1975). Amazon.com http://amzn.to/1dlExV0

International Community for Alien Research, ICAR, http://bit.ly/1aEN6v7.

Jacobs, David M. Ph.D. *The Threat*. Simon & Schuster, 1998.

Julien, Eric. *The Science of Extraterrestrials*. Fort Oglethorpe, GA: Allies Books, 2006.

Jung, Carl G. "Dr. Carl Jung on Unidentified Flying Objects." *Flying Saucer Review*, vol. 1, no. 2, 1955.

Justice, J. *DNA in the Sands of Time*. Pittsburgh, PA: Rosedog Books, 2006.

Kaku, Michio, "The physics of extraterrestrial civilizations: How advanced could they possibly be?" *www.mkaku.org/articles/physics_of_alien_civs.shtml*, 2005.

Kanazawa, Satoshi. "Do extraordinary claims require extraordinary evidence?" *http://bit.ly/1chFcqK*.

Kardashev, Nikolai. "Transmission of information by extraterrestrial civilizations." *Soviet Astron*. 8 (1964), 217. Available as a PDF file.

Kardec, Allan. *The Spirits Book: 1019 Questions and Answers about the Immortality of the Soul*. Guilford, UK: White Crow, 2009. Originally published, 1857.

Kean, Leslie. *UFOs: Generals, Pilots, and Government Officials Go on the Record*. Amazon Kindle, 2010.

Kitei, Lynne D., M.D. *The Phoenix Lights*. Charlottesville, VA: Hampton Roads 2004.

Kling, Peter. *Letters to Earth*. Durham, CT: Eloquent Books, 2009.

Knight-Jadczyk, Laura. *High Strangeness: Hyperdimensions and the Process of Alien Abduction*. Grand Prairie, Alberta: Red Pill Press, 2008.

Komarek, Ed. *UFOs, Exopolitics, and the New World Disorder*. Cairo, GA: Shoestring Publishing, 2012.

Kramer, Miriam. "Scale of universe measured with 1 percent accuracy," *http://bit.ly/1e96Rhq*.

Krapf, Philip H. *The Contact Has Begun*. Carlsbad, CA: Hay House, 2001.

———. *Meetings with Paul: An Atheist Discovers His Guardian Angel*. San Rafael, CA: Origin Press, 2008.

Krauss, Lawrence M. *A Universe from Nothing*. Simon & Schuster, 2012.

Kubris, P., and M. Macy. *Conversations Beyond the Light with Departed Friends and Colleagues by Electronic Means.* Boulder, CO: Griffin Publishing, 1995.

Lamiroy, Manuel S. *Exopaedia.* www.exopaedia.org.

LaViolette, Paul A., Ph.D. *Decoding the Message of the Pulsars.* Rochester, VT: Bear & Co., 2006.

———. *Secrets of Antigravity Propulsion.* Rochester, VT: Bear & Co., 2008.

———. *The Talk of the Galaxy.* Alexandria, VA: Starlane Publications, 2000.

Lazlo, "An unexplored domain of nonlocality: Toward a scientific explanation of Instrumental Transcommunication." http://bit.ly/1cTzBsY.

Ledwith, Miceal, D.D., Ph.D., and Klaus Heinemann, Ph.D. *The Orb Project.* Atria Books, 2007.

Linde, Andrei. "Inflation, quantum cosmology, and the anthropic principle." In J. D. Barrow, P.C.W. Davies, and C. L. Harper, eds., *Science and Ultimate Reality: From Quantum to Cosmos* (Cambridge University Press, 2003). This article is available online at arXiv.org/abs/hep-th/0211048. The author's website has more information at *www.hep.upenn.edu/~max/multiverse.ht.*

Linde, Andrei, and Vitaly Vanchurin, "How many universes are in the multiverse?" *Phys.Rev.* D81:083525, 2010. http://arxiv.org/abs/0910.1589.

Lipton, Bruce H., Ph.D. *The Biology of Belief.* Carlsbad, CA: Hay House, Kindle Edition, 2008.

Lloyd, Andy. *Dark Star.* Santa Barbara, CA: Timeless Voyager Press, 2005.

Lobuono, George. *Alien Mind: The Thought and Behavior of Extraterrestrials.* Davis, CA: QC Press, 2010.

Lora, Doris, and Russell Targ. "How I was a psychic spy for the CIA and found God." Okland, CA: Institute for Noetic Sciences (IONS), November 2003.

Lorgen, Eve. *The Love Bite: Alien Interference in Human Love Relationships.* Bonsall, CA: E Logos, 1999.

Love, Frank. *The Broken Code.* Bloomington, IN: Author House, 2007.

Mack, John E. *Passport to the Cosmos: Human Transformation and Alien Encounters.* Crown, 1999.

Maddaloni, Vincenzo, "Invention of a machine for photographing the past." Originally published in Italian in *Domenica del Corriere*, 74, no.18 (May 2, 1972), 26-29. Mark D. Williams, trans. Andrew D. Basiago, ed. http://on.fb.me/HykL0u.

Manning, Jeane, and Joel Garbon, *Breakthrough Power*. Vancouver, BC: Ambre Bridge, 2009.

Mannion, Michael, *Project Mindshift: The Re-education of the American Public Concerning Extraterrestrial Life*. Evans, 1998.

Markowitz, Clara. "Five reasons we may live in a multiverse." Science.com, December 7, 2012. http://bit.ly/18Q7t8N.

Marrs, Jim. *Alien Agenda: Investigating the Extraterrestrial Presence Among Us*. Harper Collins, 2000.

———. *Psi Spies*. Phoenix, AZ: Alien Zoo Publishers, 2000.

Mars Anomaly Research Society (MARS). www.projectmars.net.

Marzulli, L.A. *Politics, Prophecy & the Supernatural*. Crane, MO: Anomalos Publishing, 2007.

McDonald, James E. "Dr. James McDonald's letter and statement to the United Nations, June 7th, 1967." www.ufoevidence.org/documents/doc1056.htm.

Meier, Eduard. *UFOs: Spaceships from Foreign Worlds*. Schmidrüti, Switzerland: FIGU, 1999.

Mendez, Bernard. Project Pegasus. http://on.fb.me/1adQyyt.

Miejan, Tim. "Journey of souls with Michael Newton." *Edge Magazine*, July 1997. http://bit.ly/1bpsY2m.

Moody, Raymond A., Jr., M.D. *Life After Life*. Harper One, 1975.

Moore, Judith K. *Song of Freedom*. Flagstaff, AZ: Light Technology Publishing, 2002.

Moroney, Jim. *The New Bridge: Planning for the Extraterrestrial Presence*. Calgary, AB: SCMC, 2007.

Munoz, Daniel. *Enigma de los Circulos de Inglaterra*. Jiutepec, Morelos, Mexico: Espacio Alternativo, 2009.

National Aeronautics and Space Administration (NASA), Mars Exploration Program, http://1.usa.gov/16wjz3Z.

———. Photojournal, PIA10214, Spirit's West Valley Panorama, http://1.usa.gov/18Ue2Iu.

———. "Scientists discover first of a new class of extrasolar planets." Houston, TX, August 31st, 2004. http://bit.ly/1dW1UZ2.

Newlove-Eriksson, Lindy, and Eriksson, Johan, "Governance beyond the global: Who controls the extraterrestrial?" *Globalizations*, vol. 10 (2013), no. 2. *http://bit.ly/1g6woaX*.

Newton, Michael, Ph.D. *Life Between Lives: Hypnotherapy for Spiritual Regression.* Woodbury, MN: Llewellyn Publications, 2004.

———. *Destiny of Souls: New Case Studies of Life Between Lives.* Woodbury, MN: Llewellyn Kindle Edition, 2009.

———. *Journey of Souls: Case Studies of Life Between Lives.* Woodbury, MN: Llewellyn 2008

Norman, Ernest L. *The Truth About Mars.* El Cajon, CA: Unarius Academy of Science, 1998.

Oakford, David L. *Journey Through The World Of Spirit, God, Gaia, And Guardian Angels.* Forest Hill, CA: Reality Entertainment, 2007.

Ocean, Joan. *Dolphin Connection.* Kailua, HI: Dolphin Connection, 1996.

Ostrander, Sheila, and Lynn Schroeder. *Psychic Discoveries Behind the Iron Curtain.* Random House, 1971.

Palacios, Rafael. *Extraterrestres: El Secreto Mejor Guardado.* Madrid, Spain: Palmyra, 2009.

Paquin, Marguerite. *Manual for the Soul.* White Pup Press, 2009. http://bit.ly/1h6lDGh.

Peake, Anthony. *Is There Life After Death?* Arcturus Publishing. Kindle Edition, 2006.

Pearse, Steve. *Set Your Phaser to Stun.* Bloomington, IN: Xlibris, 2011.

Penrose, Roger. *The Emperor's New Mind.* Oxford: Oxford University Press, 1989.

———. *Shadows of the Mind.* Oxford: Oxford University Press, 1994.

——— and S.R. Hameroff. "Orchestrated reduction of quantum coherence in brain microtubules: A model for consciousness." In S.R. Hameroff, A.W. Kaszniak, and A.C. Scott, eds., *Toward a Science of Conscious-*

ness: The First Tucson Discussions and Debates (Cambridge, MA: MIT Press, 1996), pp. 507-540.

Pereira, Patricia. *Arcturan Star Chronicles*. Hillsboro, OR: Beyond Words Publishing, 4 vols., 1998.

Pinchbeck, Daniel. *2012: The Return of Questzalcoatl*. Tarcher/Penguin, 2006.

Proud, Louis. *The Secret Influence of the Moon*. Rochester, VT: Destiny Books, 2013.

Puthoff, Hal E. CIA-Initiated Remote Viewing at Stanford Research Institute." Austin, TX: Institute for Advanced Studies, 1995.

Rees, Martin. *Our Cosmic Habitat*. Princeton, NJ: Princeton University Press, 2001.

Reincarnation Research, http://bit.ly/HKqjF9.

Relfe, Stephanie, B.Sc. *The Mars Records*, 2 vols. www.themarsrecords.com. http://bit.ly/1aUt8Ln.

Religa, Stella, and Byron Belitsos. *The Secret Revelation: Unveiling the Mystery of the Book of Revelation*. Novato, CA: Origin Press, 2002.

Report of the Scientific Advisory Panel on Unidentified Flying Objects Convened by the Office of Scientific Intelligence, CIA, January 14-18, 1953. The Durant report of the Robertson Panel proceedings. http://bit.ly/15ry5Na.

Reynolds, C. F. *Invoking the Light*. Maryborough, Victoria, Australia: McPherson's Printing Group, 2007.

Rhine, J. B. *Extra-Sensory Perception*. Boston, MA: Bruce Humphries, 1934.

— — —. *New World of the Mind*. William Sloane, 1953.

— — —, ed. *Progress in Parapsychology*. Durham, NC: Parapsychology Press 1971.

— — —. See also http://bit.ly/HbhdBN.

Rhine, J. B., and R. Brier, eds. *Parapsychology Today*. Citadel, 1968.

Rhine, J. B., and J.G. Pratt. *Parapsychology: Frontier Science of the Mind*. Springfield, IL: Charles C. Thomas, 1957.

Ring, Kenneth, Ph.D. *Lessons from the Light: What We Can Learn from Near-Death Experiences*. Needham, MA: Moment Point Press, 1998, 2006.

Risi, Armin. *TranscEnding the Global Power Game*. Flagstaff, AZ: Light Technology Publishing, 2003. Amazon.com, http://amzn.to/Kv0Dh9.

Robinson, James, *et al.*, eds. and trans. *The Nag Hammadi Library in English*. Harper, 3d ed., 1989.

Rocky Mountain Conference on UFO Investigation. University of Wyoming, June 2000. http://bit.ly/1fM1yHn.

Rodwell, Mary, Director, ACERN. "Extraterrestrials, human consciousness and dimensions of soul: The intimate connection." http://bit.ly/19TrVWY.

Roper Poll. *UFOs and Extraterrestrial Life: Americans' Beliefs and Personal Experiences*, prepared for the SCI-FI Channel, September 2002.

Rosen-Bizberg, Franz. *Orion Transmissions Prophecy, Vol. I*. Poland: Fundacja Terapia Homa, 2003. Amazon.com, http://amzn.to/1dWaYxd.

Salla, Michael E., Ph.D. "Critique of Stanley A. Fulham case," formerly at http://bit.ly/1ajXXbA.

— — —. "A report on the motivations and activities of extraterrestrial races: A typology of the most significant extraterrestrial races interacting with humanity." Exopolitics.org, January 1, 2005. http://bit.ly/H0ajyp.

— — —. *Exopolitics: Political Implications of the Extraterrestrial Presence*. Tempe, AZ: Dandelion Books, 2003.

— — —. *Exposing U.S. Government Policies on Extraterrestrial Life: The Challenge of Exopolitics*. Amazon Kindle, 2012.

— — —. *Galactic Diplomacy: Getting to Yes with ET*. Amazon Kindle, 2014.

— — —. *Kennedy's Last Stand: Eisenhower, UFOs, MJ-12 & JFK's Assassination*. Amazon Kindle, 2013.

Sambhava, Padma, and Robert Thurman. The Tibetan Book of The Dead. Viking, 2006.

Sauder, Richard, Ph.D. *The Richard Sauder Briefing*. Amazon.com, 2010. http://amzn.to/1jkcw63.

Schuessler, John F. "UFOs, the UN, and GA 33/426." HUFON Report Newsletter, December 1992, at www.ufoevidence.org/documents/doc748.htm.

Scientific American. *Possibilities in parallel: Seeking the multiverse.* Scientific American Books, May 20, 2013. *http://books.scientificamerican.com/sa-ebooks/books/possibilities-in-parallel-seeking-the-multiverse/.*

Scott, A. *Stairway to the Mind: The Controversial New Science of Consciousness.* Copernicus / Springer-Verlag, 1995.

Seife, Charles. *Zero: The Biography of a Dangerous Idea.* London: Creative Print & Design, 2000.

Silva, Freddy. Secrets of the Fields: The Science and Mysticism of Crop Circles. Hampton Roads, VA: Hampton Roads Publishing Co. 2002.

Sivananda, Swami. *Mind: Its Mysteries and Control.* Amazon.com, 2009. http://amzn.to/1f3AvUh.

Slick, Matt. "Extraordinary claims require extraordinary evidence." Carm.org. http://bit.ly/1aAH796.

Smith, Angela Thompson, Ph.D. "The Exobiology project: Talking with the Small Greys," January 17th, 2006. Introduction, Exobiology – Small Greys – Part 1. http://bit.ly/18MG4D3.

— — —."The Exobiology project: Talking with the Hooded Reptilian," January 28th, 2006. Exobiology – Hooded Reptilian – Part 1. [Private email, not yet online.]

Smith, Jerry E., and George Piccard. *Secrets of the Holy Lance.* Kempton, IL: Adventures Unlimited Press, 2005.

Smithsonian Institution. "What does it mean to be human?" http://bit.ly/GYzJfW.

Sparks, Jim. *The Keepers.* Columbus, NC: Wildflower Press, 2006.

Stanley, Robert M. *Covert Encounters in Washington, DC.* Providence, RI: Unicus Press, 2011.

Stapp, Henry. "The Copenhagen interpretation." *Amer. J.Phys.*, 40 (1972), 1098-1116.

— — —. *Mind, Matter, and Quantum Mechanics.* Springer, 1993.

— — —. "Attention, intention, and will in quantum physics." *J. Consc. Stud.*, vol. 6 (May 24, 1999).

Steiger, Brad. *Guardian Angels and Spirit Guides.* Signet Visions,1995.

Steinberg, Max, Ph.D. *Celestial Science.* Steinberg, 2011. Scribd.com, http://bit.ly/1idtcOX.

Stevenson, Ian. "A new look at maternal impressions: An analysis of 50 published cases and reports of two recent examples." *J. Scientific Exploration*, 6 (1992), 353-373.

———. "Phobias in children who claim to remember previous lives." *J. Scientific Exploration*, 4 (1990), 243-254.

———. "American children who claim to remember previous lives." *J. Nervous and Mental Disease*, 17 (1983), no. 1, 742-748.

———. *Cases of the Reincarnation Type, I: Ten Cases in India*. Charlottesville: University Press of Virginia, 1975.

———. *Children Who Remember Previous Lives*. Charlottesville: University Press of Virginia, 1987.

———. *Science, the Self, and Survival After Death: Selected Writings*. Emily Williams Kelly, ed. Rowman and Littlefield, 2013.

Streiber, Whitley. *Transformation: The Breakthrough*. William Morrow, Beech Tree Books, 1988.

Sturrock, Peter A. *The UFO Enigma*. Warner Books, 1999.

Susskind. Leonard. *The Cosmic Landscape: String Theory and the Illusion of Intelligent Design*. Back Bay Books, 2006. Amazon.com, http://amzn.to/1bccwQ4.

Talbot, Michael. *The Holographic Universe*. Harper Perennial, 1991.

Targ, Russell. *The Reality of ESP: A Physicist's Proof of Psychic Abilities*. Quest Books, 2012.

Targ, Russell, and Harold Puthoff, Ph.D. *Mind-Reach*. Charlottesville, VA: Hampton Roads, 1977.

Targ, Russell, and Katra, Jane. *The Scientific and Spiritual Implications of Psychic Abilities*. Palo Alto, CA. www.espresearch.com/espgeneral/doc-AT.shtml, 2005.

Taylor, Greg, "Co-Founder of String Field Theory Explores the Physics of ET." www.space.com, October 30th, 2003.

Tegmark, Max. "Is 'The Theory of Everything' merely the ultimate ensemble theory?" *Annals Phys.*, 270, no.1 (November 20, 1998), 1–51. Available online at arXiv.org/abs/gr-qc/9704009.

Tellinger, Michael, and Johan Heine. *Temples of the African Gods*. Waterval Boven, South Africa: Zulu Planet, 2009.

Tellinger, Michael. *Slave Species of God*. Johannesburg, South Africa: Zulu Planet, 2009.

Torres, Noe, and Ruben Uriarte. *Mexico's Roswell: The Chihuahua UFO Crash*. Roswell Books 2008. Amazon.com, http://amzn.to/1eXwz9J.

Tramont, C.V. M.D. *From Birth to Rebirth*. Columbus, NC: Swan-Haven. Amazon.com, http://amzn.to/1fOiFte.

Tremblay, Rodrigue. *The Code for Global Ethics*. Amherst, NY: Prometheus Books, 2010.

Tunich, Jose Alberto. *Encuentros Cercanos con Extraterrestres*. Buenos Aires, Argentina: CEFAE, 1999. http://www.fuerzaaerea.mil.ar/prensa/cefae.html.

UFO evidence. http://www.ufoevidence.org/topics/GeneralSightings.htm.

United Nations General Assembly, Recommendation to Establish UN Agency for UFO Research – UN General Assembly decision 33/426, 1978.

United Nations Treaty on Principles Governing the Activities of States in the Exploration and Use of Outer Space, including the Moon and Other Celestial Bodies, London, Moscow, and Washington, January 27th, 1967. www.oosa.unvienna.org/SpaceLaw/ outerspt.html.

University of Virginia Medical School, Division of Perceptual Studies. http://bit.ly/17cJkMo.

Urantia Foundation. *The Urantia Book*. http://bit.ly/1fCFBIr.

Vasquez, John, and Bruce Stephen Holms. *Incident at Fort Benning*. Santa Barbara, CA: Timeless Voyager Press, 2000.

Vaughan-Lee, Llewellyn. *Awakening the World*. Inverness, CA: Golden Sufi Center, 2006.

Vilenkin, Alexander. *Many Worlds in One: The Search for Other Universes*. Hill and Wang, 2006.

Vilenkin, Alexander, and Max Tegmark. "The case for parallel universes." *Scientific American*, July 11, 2011. http://www.scientificamerican.com/article.cfm?id=multiverse-the-case-for-parallel-universe.

Von Neumann, John. *Mathematical Foundations of Quantum Mechanics*. Princeton, NJ: Princeton University Press, 1955.

Walia, Arjun "Russian prime minister confirms the existence of intelligent extraterrestrial life." http://bit.ly/1hYqkT4.

Wallace, Tim. *Hidden Wisdom*. Newburyport, MA: Disinformation Books, 2010. http://bit.ly/1dKTJ2J.

Ward, Suzanne. *Matthew: Tell Me About Heaven: A Firsthand Description of the Afterlife*. Camas, WA: Matthew Books, 2006.

Webre, Alfred Lambremont. "A 'New Cydonia' of ancient extraterrestrial monuments found on Mars." http://bit.ly/1cdAGeS.

———. "Andrew Basiago is predicted 'planetary level' whistleblower for Mars life and time travel." *http://exm.nr/1c5NUdW*; http://bit.ly/19NywlH

———. "Archons: Exorcising hidden controllers with Robert Stanley and Laura Eisenhower." *http://bit.ly/1crsAU1*; http://exm.nr/HfQpQj.

———. "Are the UFOs appearing in Russia skies in Dec. 2010 Fulham's predicted UFO wave?" http://exm.nr/17y2DQW.

———. "Secret US Mars program and life on Mars." http://bit.ly/1g7ywC6.

———. "Basiago and Eisenhower reveal 'Marsgate' and make case for 'Alternative 4'." *http://bit.ly/17oh5dZ*; http://exm.nr/17ya8VD.

———. "Discovery of life on Mars by Andrew D. Basiago chosen #1 UFO story of 2008." http://bit.ly/1akYe3U.

———. "ET council seeded *Homo sapiens* as intelligent beings with 12-strand DNA." http://exm.nr/1h6KiOF.

———. "ET council: We will increase UFOs, address U.N. in 2014, renew ecology in 2015." *http://exm.nr/HtiNyQ*.

———. Exopolitica: La Politica, El Gobierno y La Ley en el Universo. Granada:Vesica Piscis, 2009. *http://amzn.to/185GYdq*.

———. *Exopolitics: A Decade of Contact*. Vancouver, BC: Universe Books, 2000.

———. *Exopolitics: Politics, Government, and Law in the Universe*. Vancouver, BC: Universe Books, 2005.

———. "Exopolitics researcher develops evidence-based typology of extraterrestrial civilizations." Examiner.com, April 12, 2010. http://exm.nr/1fy4bwn

———. "EXOPOLITICS: The discovery of life on Mars with Andrew D. Basiago." *http://bit.ly/1dEI8DD*.

———. "GOOD NEWS! Universe singularity now emanating pre-wave energy for 'enlightened unity consciousness'." *http://exm.nr/19ZSytf*; http://bit.ly/1avePQ0.

———. "How intelligence legend and Manchurian candidate Barack Hussein Obama was created." http://slidesha.re/18Im2fE.

———. "Human souls common to extraterrestrials and earthlings, researcher says." http://bit.ly/18mlpLv.

———. "Informal UFO/ET disclosure now happening in leaked UFO/ET reports via US Capitol Police." http://exm.nr/17zQyuo.

———. "Intention experiments, interdimensional UFOs, and multidimensional beings converge at Mt. Adams, WA." http://exm.nr/18DT1gZ.

———. "Is UFO orb over Dome of the Rock in Jerusalem a context communication by ET?" http://exm.nr/HjWICv.

———. "Israeli media: 'Jerusalem UFO orbs fulfill Stan Fulham's ET council prediction'." http://exm.nr/19XKmK4.

———. "January 2011 UFO wave over Moscow is 3rd independent confirmation of ET council." http://exm.nr/1iB7osb .

———. "Mars active industrial site located by remote viewing, JPL photos, corroborating Mars whistleblowers." http://bit.ly/17sIc5H.

———. "MARS takes its case for life on Mars to the American people." http://bit.ly/1gVXCFp.

———. "Mars visitors Basiago and Stillings confirm Barack Obama traveled to Mars." http://bit.ly/1OL2P5t.

———. "Mary Rodwell – ETs, souls, the new humans and a coming global shift: Interview." http://bit.ly/17tBY7B.

———. "More predicted UFO sightings over New York confirm ET will intervene in ecology." *http://exm.nr/17CmVq5*.

———. "Moscow Jan. 2011 UFO wave is ongoing, putting Fulham prediction critics to shame." http://exm.nr/1atc6co.

———. "My 1970s meeting with DARPA's Project Pegasus secret time-travel program." http://bit.ly/182i3dE

———. "Mystery deepens around Norway spiral light: Pyramid UFO over Kremlin, and Russia and China lights." http://exm.nr/1aSi0lF.

———. "New data, law of evidence support view of Mars having indigenous, intelligent extraterrestrial life." http://bit.ly/P9nxw6.

———. "NZ UFO expert describes Grey ET management of human souls and bodies aboard Grey ET spacecraft." http://bit.ly/17Y7MSq.

———. "NZ UFO expert: On a UFO, Grey ETs merged my son's soul into his body in my womb (photos)." http://exm.nr/1eWlXJF.

———. "Occupy Adam's Calendar, Part I, Extraterrestrial genetic manipulation: Geneticist William Brown" (film). http://bit.ly/W9o4jH.

———. "Part 2 - Peter Kling: Christ returns as an extraterrestrial with the armies of the multiverse to defeat the NWO?" *http://bit.ly/JSviEI.*

———. "Report: Between 2 percent of US population and 14.7 percent of global population are being involuntarily teleported (abducted) by hyperdimensional civilizations." http://bit.ly/1f4GcCi.

———. "Second whistleblower emerges to confirm reality of time travel." *http://exm.nr/18T86P5.*

———. "Spectacular Jan 18, 2011, Moscow UFO part fulfillment of ET council's prediction?" http://exm.nr/1hAl7Sd.

———. "Stanley Fulham dies, warned ETs will intervene and save Earth's collapsing ecology." http://exm.nr/16T0j6m.

———. *The Age of Cataclysm.* G.P. Putnam's Sons, 1974; Berkeley Medallion, 1975; Capricorn Books, 1975. Tokyo: Ugaku Sha, 1975.

———. *The Levesque Cases.* Ontario: PSP Books, 1990.

———. "Third whistle blower confirms Obama's participation in CIA jumproom program of early 1980's." http://bit.ly/O5TRcO

———. "Time travel and political control." http://bit.ly/LoBF0T

———. "Two whistleblowers independently report teleporting to Mars and meeting Martian extraterrestrials." http://bit.ly/1hnCcyQ.

———. "UFO expert Stanley Fulham UFO predictions fulfilled over NY, Moscow, and London." http://exm.nr/1ghJJ39.

———. "UFO near White House emits ray of light: Horseplay, false flag or 'socially destabilizing' event?" http://exm.nr/1aSMOmm.

———. "Up to 1 billion humans are abducted by hyperdimensional ETs, and humans are in cognitive dissonance," http://exm.nr/HfEpOC.

———. "Vancouver futurist and former DARPA time traveler question de facto Mars truth embargo." http://bit.ly/1bw5bvp.

———. "When will extraterrestrial 'disclosure' or contact in the public domain happen?" http://bit.ly/1gdjmLC.

———. "Whistleblower exposes attempted ET manipulation, false flag at 'Festival'." http://exm.nr/17zzQLM.

———. "Whistleblower Laura Magdalene Eisenhower, Ike's great-granddaughter, outs secret Mars colony project." http://bit.ly/1aPgkG8.

———. "Why is Hawking affirming time-travel theory and appearing ignorant of DARPA secret time travel?" *http://exm.nr/GO2Cvz*.

———. www.exopolitics.com.

Wetunde, Ahaze. "What are rules governing the admissibility of evidence in court?" http://bit.ly/1cRN8Tk.

Wilber, Ken. A Brief History of Everything. Boston, MA: Shamballa, 2007.

Wilber, Ken. Integral Spirituality: A Startling New Role for Religion in the Modern and Post Modern World. Boston, MA: Integral Books, 2007.

Wilcock, David. *The Source Field Investigations: The Hidden Science and Lost Civilizations Behind the 2012 Prophecies.* Amazon Kindle, 2012.

———. *The Synchronicity Key: The Hidden Intelligence Guiding the Universe and You.* Penguin Group US Kindle Edition, 2013.

Wilson, Simon Harvey. "Shamanism and alien abductions: A comparative study."*Australian J. Parapsychol.*, vol. 1, no. 2 (December 2001). http://bit.ly/1h6tMu9.

Wolchover, Natalie, and *Quanta Magazine*. "New physics complications lend support to multiverse hhypothesis." *Scientific American*, June 1, 2013. *http://www.scientificamerican.com/article.cfm?id=new-physics-complications-lend-support-to-multiverse-hypothesis*.

Wood, Ryan S. *Majic Eyes Only: Earth's Encounters with Extraterrestrial Technology*. Wood Enterprises 2005. Amazon.com, http://amzn.to/1dHCqRH.

Woolf, Victor Vernon. *The Dance of Life*. Reno, NV: International Academy of Holodynamics, 2005.

Wyllie, Timothy. *Revolt of the Rebel Angels: The Future of the Multiverse.* Rochester, VT: Bear & Co., 2013.

Zammit, Victor, and Wendy Zammit. *A Lawyer Presents the Evidence for the Afterlife.* White Crow Books Kindle Edition, 2013.

Padma Sambhava, Robert Thurman. *The Tibetan Book of The Dead.* New York, NY: Viking, 2006.

Wilber, Ken. *A Brief History of Everything.* Boston, MA: Shamballa, 2007.

Wilber, Ken. *Integral Spirituality: A Startling New Role for Religion in the Modern and Post Modern World.* Boston, MA: Integral Books, 2007.

INDEX

abduction, abductees, ix, 25, 28, 42, 43, 45, 71, 83, 84, 85, 95, 165, 170, 171, 180

ACERN, 95, 165, 185, 197

afterlife, vi, vii, viii, ix, xii, xvi, 6, 7, 8, 9, 11, 12, 18, 23, 25, 31, 32, 124, 125, 132, 137, 167, 172

Alpha Centauri, 108

American Astronomical Society, 5

Amoroso, Richard L., 3, 4, 185

archeology, 28

Archons, 93, 94, 201

Arcturian humans, 21

Area 51, 72

Arrigo, Jean Maria, 57

Australian Close Encounter Research Network, 95, 185

Badme, Mijaliáyev, 112

Barcelona Science and Spirit Conference, 63

Baryon Oscillation Spectroscopic Survey, 5

Basiago, Andrew D., 18, 19, 20, 40, 54, 55, 56, 57, 58, 59, 60, 61, 62, 63, 64, 65, 66, 67, 68, 69, 70, 71, 72, 73, 74, 75, 76, 77, 87, 88, 119, 129, 149, 150, 185, 189, 194, 201, 202

Beamer, Dave, 56

Bielek, Al, 90

Biocentrism, 140, 186

Bootes, 108

Bromberg, Facundo, 15, 186

California Institute of Technology, 69

Calleman, Carl Johan, 178, 179, 186

Carter, Chris, 19, 30, 176, 186, 187, 217

Central Intelligence Agency (CIA), 7, 19, 55, 56, 57, 58, 59, 60, 61, 65, 66, 67, 68, 70, 71, 72, 73, 74, 76, 77, 78, 79, 176, 193, 196, 203, 217

channelers, 28

chrononaut, 19, 40, 54, 57, 58, 59, 65, 72, 87, 129, 149

chronovision, xvi, 128, 130

chronovisor, 19, 129, 130

civilizations, iii, v, viii, ix, x, xi, xii, xiii, xiv, xv, xvi, xvii, xviii, xix, xx, 6, 7, 9, 12, 13, 16, 17, 18, 26, 27, 29, 30, 31, 34, 35, 36, 37, 38, 46, 47, 51, 54, 76, 81, 82, 83, 85, 87, 88, 92, 95, 101, 102, 104, 105, 106, 107, 108, 117, 119, 123, 132, 142, 143, 145, 150, 152, 154, 159, 160, 163, 172, 173, 175, 176, 178, 180, 181, 182, 189, 192, 201, 203, 217

Close Encounters of the Third Kind, 46

Clow, Barbara Hand, 1, 21, 36, 187

207

Clow, Gerri, 1, 21, 36, 187

Cognitive dissonance, 84

College of the Siskiyous, 66, 67, 68, 73, 74

Columbia University, 67

Comsuli, 108

consciousness, x, xiv, xv, xix, 11, 15, 20, 21, 22, 23, 24, 25, 34, 35, 36, 38, 47, 82, 83, 90, 93, 94, 95, 96, 97, 99, 100, 103, 105, 109, 123, 127, 133, 140, 141, 148, 156, 159, 176, 177, 178, 179, 180, 189, 195, 197, 202, 218

conspiracy of silence, 110, 112, 113, 114

contact, ix, 24, 33, 46, 47, 56, 77, 85, 98, 102, 105, 106, 111, 116, 126, 127, 128, 140, 141, 156, 159, 204

contactees, viii, 28, 83, 95, 119, 180

Corkscrew, 56, 58, 76

Cowen, Ron, 18

cross-correspondences, 139

Curtiss-Wright Aeronautical Company, 19, 57

Dames, Ed, 59, 65, 67, 68, 69, 73, 74

DARPA, 18, 19, 55, 57, 59, 66, 73, 77, 202, 204, 217

Dean, Robert, 68

Defense Advanced Research Projects Agency, 66, 70, 217

density, x, 22, 23, 24, 25, 34, 35, 38, 150, 155

density-based models, 23, 24

Depanoid, 45

dimensional portal, 18, 179

dimensionality, 1, 177

disinformation, 7, 29, 115, 117, 119

displays, 109, 114, 117, 118

Dreamweaver, 156

Dugan, Regina E., 59, 66, 67, 68, 69, 72, 73

Dunham, Stanley A., 59, 66

eclipses, 116

Eisenhower, 59, 70, 73, 88, 93, 94, 189, 197, 201, 204

Electronic Voice Phenomena, xvi, 128, 142

energy frequency, x

Environmental Protection Administration, 19, 217

Ernetti, Pellegrino, xvi, 128, 129, 130, 131

evidence, v, vi, vii, viii, x, xiii, xiv, xv, xvi, xvii, xviii, 6, 7, 9, 11, 12, 14, 15, 16, 22, 23, 26, 27, 28, 31, 32, 33, 34, 36, 47, 51, 52, 53, 54, 55, 57, 60, 63, 64, 77, 78, 81, 82, 85, 86, 92, 94, 95, 101, 102, 105, 106, 107, 112, 117, 118, 123, 124, 125, 131, 132, 133, 134, 135, 137, 139, 141, 142, 143, 149, 159, 160, 163, 165, 172, 175, 181, 182, 185, 192, 198, 200, 201, 203, 204

Exobiology Project, 41, 42, 44

Exophenotypes, xiii, xiv, 46, 47, 48, 51, 55, 56, 57, 61, 62, 63, 64, 185

Exophenotypology, xiii, xiv, 38, 40, 42, 45, 54, 55, 60

Exopolitics, 7, vi, vii, xi, xv, xix, 11, 13, 19, 26, 27, 28, 30, 31, 32, 34, 38, 52, 54, 68, 102, 110, 171, 172, 176, 191, 192, 197, 201, 217

extraterrestrial and hyperdimensional, xi, 8

Extraterrestrial Communication Study, 19

Extraterrestrials, 39, 46, 95, 100, 102, 159, 171, 188, 191, 192, 193, 197

False Flag operation, 85, 92, 115, 118

Farsight Institute, 103, 104

Fontana, David, 137, 189

Freer, Neil, 83

Frei, Gebhard, 129

Fulham, Stanley A., 107, 108, 109, 110, 111, 112, 113, 114, 115, 116, 117, 118, 119, 189, 197, 201, 202, 203

galaxy, galaxies, xi, xiv, xv, 4, 5, 6, 7, 21, 22, 35, 36, 37, 44, 82, 86, 89, 92, 101, 104, 106, 107, 163, 172, 178, 179

Garcia, Ernest, 57

Gemelli, Augustino, xvi, 128, 129, 130, 131

Giant snakes, 44

Goswami, Amit, 14, 15, 16, 186, 189, 190

Grey extraterrestrials, xviii, 74, 75, 76, 78, 88, 157, 166, 174

Greys, Standard, 40, 41, 42, 43, 44, 56, 71, 72, 75, 76, 77, 78, 83, 89, 98, 100, 166, 168, 169, 180, 198

Gurdjieff, George, 22, 25

Gurney, Edmund, 139

Gusev Crater, 62

Hameroff, Stuart, 141, 195

Hansen, Suzanne, xviii, 83, 164, 165, 166, 167, 168, 169, 170, 171, 173

Haramein, Nassim, 179

Heim, Burkhard, 127

Hohenwarter, Peter, 129

holographic fragment, xvi, 7, 17, 32, 124, 160, 181

Hopkins, Budd, 83

Hoyle, Sir Fred, x

Hughes Aircraft, 66

humanoid, xiii, 39, 40, 45, 46, 51, 52, 55, 56, 58, 61, 62, 64, 185

Hunt, 56, 58, 59, 67, 68, 71, 76, 188

Huyghe, Patrick, 38, 46, 191

hybrid, 45, 167

Hynek, J. Allen, 46, 191

hyperdimensional, iii, ix, xiii, xiv, xvii, 30, 40, 44, 47, 51, 81, 82, 83, 84, 85, 86, 87, 88, 89, 91, 92, 93, 95, 98, 100, 106, 108, 114, 115, 119, 158, 159, 160, 163, 164, 167, 168, 169, 170, 174, 178, 180, 181, 203

hypnotic regression, xvi, 6, 23, 84, 99, 124, 132, 143, 144, 147, 148, 149, 154, 155, 164, 167, 171, 172, 173, 177

hypothesis, xii, xiii, xv, xvi, xviii, xix, xx, 3, 7, 12, 13, 14, 16, 20, 22, 25, 31, 32, 51, 81, 95, 100, 104, 115, 118, 123, 128, 131, 132, 133, 137, 143, 149, 152, 154, 155, 160, 163, 171, 172, 173, 175, 180, 181, 182, 191, 204, 217

incarnation, xviii, xix, xx, 97, 99, 133, 146, 147, 159, 164, 173, 177, 181

indigenous intelligent life on, 52

infinite universe, 5, 14

insectoid, 45

Instrumental Transcommunication, xvi, 124, 125, 126, 130, 131, 142, 193

intelligences, 33, 34, 95, 96, 99, 100, 103, 124, 130, 159, 165, 170 intelligent, v, viii, ix, x, xi, xii, xiii, xiv, xv, xvi, xvii, xviii, xix, xx, 6, 7, 9, 12, 13, 14, 16, 17, 18, 26, 27, 29, 30, 31, 34, 35, 36, 37, 38, 46, 48, 51, 52, 54, 55, 60, 61, 64, 76, 81, 82, 85, 87, 88, 92, 95, 101, 102, 104, 105, 106, 107, 108, 111, 115, 116, 117, 119, 123, 124, 132, 142, 143, 144, 145, 146, 150, 151, 152, 154, 158, 159, 160, 163, 171, 172, 173, 175, 176, 177, 180, 181, 182, 201, 203, 217, 218

interdimensional, ix, x, xiv, xvii, xviii, 13, 18, 27, 29, 30, 32, 34, 47, 76, 81, 82, 93, 106, 109, 113, 115, 116, 117, 118, 131, 146, 147, 150, 155, 156, 160, 164, 170, 173, 177, 178, 202

interdimensional portal, ix, 18, 32, 131, 146, 147, 150, 160, 164, 178

Interlife, i, viii, ix, xv, xvi, xvii, xix, xx, 6, 7, 9, 11, 12, 18, 21, 23, 25, 30, 31, 32, 123, 124, 125, 128, 130, 131, 132, 137, 139, 142, 143, 144, 145, 146, 147, 148, 149, 150, 154, 155, 157, 158, 160, 164, 167, 171, 172, 173, 176, 177, 180, 182, 217

International Community for Alien Research, 83, 174, 191

International Society for Catholic Parapsychologists, 129

intervention, xviii, 108, 109, 110, 111, 112, 115, 116, 117, 119, 159, 174, 182

Jacobs, David., 83, 192

Jet Propulsion Laboratory, 69

jump room, 74, 75, 76, 77

Jürgenson, Friedrich, 129

Kaku, Michio, 35, 192

Kardashev scale, 35, 189, 190

Kardashev, Nikolai, 35, 189, 190, 192

Kitchur, Randy, 117, 118

Krauss, Lawrence M., 14, 192

La Vanguardia, vii

Lamiroy, Manuel S., 23, 38, 39, 40, 41, 42, 44, 45, 46, 47, 193

Lanza, Robert., 140, 141, 186

Lash, John, 94

Laszlo, Ervin, 125

Lawrence Berkeley National Laboratory, 5

Lawrence Livermore National Laboratory, 70

life after death, xii, 30, 32, 136, 182

Light, xvi, 7, 21, 124, 137, 144, 145, 167, 168, 170, 186, 187, 189, 190, 191, 193, 194, 196, 197

Linde, Andrei, 4, 5, 6, 29, 30, 101, 193

Lockheed Corporation, 66

Macy, Mark, 124, 129, 193

Mars, xiii, xiv, 18, 37, 51, 52, 54, 55, 56, 57, 58, 59, 60, 61, 62, 63, 64, 65, 66, 67, 68, 69, 70, 71, 72, 75, 76, 77, 81, 86, 87, 88, 89, 90, 92, 172, 185, 186, 194, 195, 196, 201, 202, 203, 204

Mars Anomaly Research Society, 61, 62, 64, 186, 194

Mars rovers, xiv, 52, 55

Martian, xiv, 40, 52, 54, 55, 56, 57, 60, 61, 62, 63, 64, 65, 67, 68, 69, 71, 76, 92, 203

Martian humanoid, xiv, 40, 55, 56, 57, 61, 62, 63, 64, 69

McCool, William C., 59, 72

mediums, ix, 126, 139, 140, 142

Mendez, Bernard, 58, 59, 60, 71, 72, 73, 74, 75, 76, 77, 78, 119, 194

Moody, Raymond A., 30, 137, 194

Mothmen, 43

multiverse, v, vii, viii, x, xi, xii, xiii, xiv, xv, xvi, xvii, xviii, xix, 3, 4, 5, 6, 7, 12, 13, 14, 17, 18, 27, 29, 30, 31, 37, 38, 51, 76, 82, 95, 96, 101, 102, 103, 117, 119, 123, 142, 144, 151, 153, 154, 157, 159, 160, 163, 164, 172, 173, 175, 177, 189, 190, 193, 194, 198, 200, 203, 204, 217

Myers, Frederick W.H., 139

National Aeronautics and Space Administration (NASA), 194

Near Death Experiences (NDE), 127, 137, 138, 139

Neumann, Arthur, 59, 70, 200

Newton, Michael, xvi, 6, 16, 18, 22, 23, 30, 32, 124, 144, 145, 147, 149, 150, 151, 152, 153, 154, 155, 156, 157, 158, 159, 160, 161, 173, 176, 177, 178, 194, 195

Non-winged Draco, 43

NORAD, 107, 111, 116, 118

number of communicating intelligent, 6, 7

Obama, Barack, 59, 65, 66, 67, 68, 69, 70, 71, 72, 73, 75, 76, 78, 79, 202, 203

Occidental College, 67, 69

of souls, v, vi, vii, viii, ix, x, xi, xii, xv, xvi, xvii, xviii, xix, xx, 7, 9, 12, 13, 16, 17, 31, 123, 124, 132, 142, 143, 144, 145, 146, 148, 150, 152, 154, 158, 159, 160, 163, 164, 173, 177, 181, 194, 217

of spiritual beings, xi, xvii, 7, 9, 13, 16, 17, 31, 143, 151, 152

Office of Naval Intelligence, 66

Omniverse, 1, 3, 5, 6, i, iii, v, vi, vii, viii, ix, x, xi, xii, xiii, xv, xvi, xvii, xviii, xix, xx, 1, 3, 12, 13, 14, 16, 17, 18, 20, 22, 23, 25, 26, 27, 28, 29, 30, 31, 32, 33, 36, 37, 49, 51, 54, 81, 95, 101, 104, 121, 123, 124, 128, 130, 132, 137, 142, 143, 149, 151, 152, 154, 160, 161, 163, 164, 175, 176, 177, 180, 181, 182, 217, 218

ontology, 26, 27, 53, 190

Orion, 108, 197

Othman, Mazian, 111

Out of Body experiences (OBE), 28, 97, 98, 99, 127

Oxford University Press, ix, 33, 34, 195

parapsychology, viii, xix, 11, 26, 30, 31, 171, 176

past lives, 132

Paul VI, 129

Peake, Anthony, 143, 195

Penrose, Roger, 141, 195

photographs, 14, 28, 45, 61, 62, 63, 64, 93, 94, 107

Pius XII, 128

Pleiadians, 72, 106, 109, 117

Pouseti, 108

Praying mantis-like, 45

principles of international law, ix, 33

Project Bluebeam, 118

Project Pegasus, 19, 55, 57, 60, 72, 77, 87, 185, 194, 202

proof, xiv, 12, 22, 53, 54, 133

psychic capacity, 30, 169

quantum mechanics, 15, 140, 190

Ralph M. Parsons Company, 57, 59, 66, 69

Rauscher, Elizabeth A., 3, 4, 185

reincarnation, vii, viii, xvii, xix, xx, 9, 11, 30, 132, 133, 135, 148, 157, 158, 159, 171, 172, 174, 176, 181, 182

Relfe, 59, 69, 70, 88, 89, 90, 92, 119, 196

religions, viii, xii, 11, 12, 32, 110, 182

remote sensing, 19

remote viewing, xv, 28, 90, 102, 103, 104, 202

reptilian, 42, 43, 44, 83, 85, 146

Rhine, J.B., 30, 196

Rodwell, Mary, 48, 82, 83, 95, 96, 98, 100, 159, 164, 165, 166, 167, 169, 170, 171, 179, 180, 197, 202

Rover, 62

Rumsfeld, Donald, 20

Rundle, Michael, 70

Sagan, Carl, xiv, 54

Sanchis, Ima, vii

Sandia National Laboratory, 19

Schlegel, David, 5

Senkowski, Ernst, 126

Sidgwick, Henry, 139

sightings, 28, 46, 107, 109, 110, 111, 112, 114, 115, 165, 202

Sirians, 106

Sirius, 108

Society for Psychical Research, 139

Soetoro, Barry, 65, 66, 67, 70, 71

solar system, xiii, xiv, 35, 36, 37, 51, 54, 56, 58, 70, 82, 85, 86, 89, 92, 152, 172, 179

Solar-system, 37, 172

soul, souls, iii, vi, vii, viii, ix, xii, xv, xvi, xvii, xviii, xix, xx, 6, 7, 9, 11, 12, 17, 18, 21, 22, 23, 24, 25, 30, 31, 32, 95, 96, 97, 98, 99, 100, 110, 123, 124, 125, 131, 132, 133, 134, 137, 139, 141, 143, 144, 145, 146, 147, 148, 149, 150, 151, 152, 153, 154, 155, 156, 157, 158, 159, 160, 163, 164, 165, 166, 167, 168, 169, 170, 171, 173, 174, 175, 176, 177, 178, 180, 181, 182, 183, 197, 202, 203

space, v, ix, x, xi, xiii, xiv, xvii, 3, 4, 5, 12, 13, 17, 18, 19, 21, 23, 29, 31, 35, 36, 37, 47, 51, 52, 54, 55, 56, 65, 69, 73, 74, 75, 76, 77, 78, 81, 82, 86, 87, 88, 89, 90, 92, 95, 96, 98, 99, 100, 102, 104, 106, 119, 126, 127, 140, 149, 152, 153, 155, 158, 159, 164, 172, 188, 199

spacecraft, ix, xviii, 107, 110, 111, 113, 114, 117, 163, 164, 165, 166, 167, 168, 171, 181, 203

Spent, Mike, 58, 59, 61

spiritual beings, v, vi, vii, viii, x, xii, xvii, xix, 12, 16, 17, 30, 31, 152, 157, 175

spiritual dimensions, v, vi, vii, viii, x, xi, xii, xv, xvi, xvii, xix, xx, 6, 7, 9, 12, 13, 17, 18, 20, 22, 23, 27, 29, 30, 31, 81, 123, 124, 128, 130, 131, 132, 139, 144, 147, 149, 152, 153, 160, 163, 171, 175, 176, 181, 183

Sprinkle, Leo, 26

Stacy, Dennis, 38, 191

Stanford Research Institute, 19, 196, 217

Stanley, Robert, 59, 66, 93, 94, 107, 108, 110, 111, 112, 113, 116, 117, 119, 189, 197, 198, 201, 203

Stargate, 28

Stevenson, Ian, 30, 132, 133, 134, 136, 199

Stillings, Brett, 56, 58, 59, 60, 65, 66, 67, 68, 69, 70, 72, 73, 74, 76, 77, 78, 119, 202

Stillings, Thomas J., 56, 58, 59, 60, 65, 66, 67, 68, 69, 70, 72, 73, 74, 76, 77, 78, 119, 202

Stillings, William B., 56, 58, 59, 60, 65, 66, 67, 68, 69, 70, 72, 73, 74, 76, 77, 78, 119, 202

Strickland, Michael, 59

string theory, xi, 4

Structural Soul, 154

synthetic quantum environment (SQE), 75, 76

Tall Greys, 40

Tall Nosed Greys, 40

Tall Winged Draco, 42

Taller humanoid, 39

Taylor, Tracey, 98, 199

telepathy, ix, xix, 11, 94, 127, 136, 176

teleportation, viii, ix, 18, 19, 20, 47, 57, 58, 59, 65, 66, 73, 74, 78, 81, 88, 95, 119, 149, 160

time loops, 20

time travel, 18, 19, 55, 57, 72, 79, 88, 89, 119, 129, 130, 201, 203, 204, 217

time-space hologram, xvi, 17, 18, 19, 87, 89, 103, 123, 124, 125, 130, 131, 133, 134, 149, 155, 160

Transcendors, 109, 115

Tsiolkovski Ridge, 56, 62

Turner, Stansfield T., 59

typology, x, xiii, xix, 28, 31, 34, 35, 36, 37, 38, 47, 54, 56, 65, 172, 177, 180, 197, 201

typology of intelligent, x, xix, 28, 31, 34, 36, 37, 38, 54, 172, 180

UFOCUS NZ Research Network, 164

UFOs, 82, 107, 108, 109, 111, 112, 114, 116, 118, 185, 186, 188, 192, 194, 197, 201, 202

United Nations, 46, 110, 111, 185, 186, 191, 194, 200, 217

universal energy wave, 179

universes, v, vi, vii, viii, x, xi, xii, xiii, xiv, xv, xvi, xvii, xviii, xx, 3, 4, 5, 6, 12, 13, 17, 18, 22, 23, 26, 27, 29, 31, 32, 33, 36, 37, 38, 81, 82, 101, 102, 104, 123, 128, 130, 140, 141, 142, 143, 145, 148, 149, 150, 152, 153, 154, 155, 158, 159, 160, 161, 163, 172, 173, 175, 181, 193, 200

universes of the multiverse, vi, vii, viii, xi, xiii, xv, xvi, xvii, xix, 6, 22, 27, 29, 31, 36, 38, 101, 140, 155, 158

Urantia book, 29

US Department of Defense, 19

Vanchurin, Vitaly, 4, 5, 6, 29, 30, 101, 193

Vietor, Tommy, 70, 71, 73, 78

vortal tunnel, 19, 149

Westrum, Ron, 83

whistleblowers, 52, 69, 70, 87, 119, 202, 203

Wilcock, David, 22, 70, 157, 204

Winged Draco, 43

World public opinion, 7

Zammit, Victor, 124, 125, 128, 129, 137, 139, 140, 205

Zeta Reticuli, 46, 108

213

About Author
Alfred Lambremont Webre

Alfred Lambremont Webre's 2000 book, *Exopolitics: A Decade of Contact*, founded Exopolitics, the science of relations among intelligent civilizations in the multiverse. Using advanced quantum access technology, the U.S. Defense Advanced Research Projects Agency (DARPA) and the Central Intelligence Agency (CIA) time traveled his 2005 book, *Exopolitics: Politics, Government, and Law in the Universe*, back to 1971, when an unwitting Alfred was examined by a group of approximately 50 CIA and DARPA officials who knew Alfred would be a leading future extraterrestrial and time-travel whistleblower, and would become the developer of the Exopolitics and Dimensional Ecology models of the Omniverse.

Alfred's new book, *The Dimensional Ecology of the Omniverse*, integrates empirical data from intelligent extraterrestrial civilizations and from the intelligent civilizations of souls in the Interlife and demonstrates a new hypothesis of a functioning ecology of intelligence in the dimensions in the Omniverse.

Alfred is a graduate of Yale University and Yale Law School in international law and was a Fulbright Scholar in international economic integration in Uruguay. He has taught economics at Yale University and constitutional law at the University of Texas. Alfred was general counsel to the New York City Environmental Protection Administration, a futurist at Stanford Research Institute (where he directed the proposed 1977 Carter White House extraterrestrial communication study), and was a NGO delegate to the United Nations and the UNISPACE conference.

Alfred is active with non-profit organizations concerned with peace, the environment, human rights, consciousness, and intelligent life in the Omniverse.

Alfred may be contacted through
www.dimensionalecology.com

Made in the USA
Lexington, KY
15 July 2014